BRIAN R. LITTLE

ME, MYSELF, AND US

The Science of Personality
and the Art of Well-Being

HarperCollins*Publishers*Ltd

Me, Myself, and Us
Copyright © 2014 by TRAQ Consulting Ltd
All rights reserved.

Published by HarperCollins Publishers Ltd

First edition

HarperCollins books may be purchased for educational, business,
or sales promotional use through our Special Markets Department.

HarperCollins Publishers Ltd
2 Bloor Street East, 20th Floor
Toronto, Ontario, Canada
M4W 1A8

www.harpercollins.ca

Library and Archives Canada Cataloguing in Publication
information is available upon request

ISBN 978-1-44340-186-9

Printed and bound in the United States of America
RRD 9 8 7 6 5 4 3 2 1

For my wife, Susan, with love

Contents

Preface

ME, *MYSELF, AND US* EXPLORES QUESTIONS THAT ARE rooted in the origins of human consciousness but are as commonplace as the conversation you had at breakfast this morning. The questions are very personal and deal with *you, yourself*. Am I really an introvert? Why can I motivate my employees but can't connect at all with my kids? Why am I a completely different person at home from the person I am at work? Do I really control the things that matter to me? I seem to be uncommonly happy—is there something wrong with me? Is there any truth to the ludicrous rumor that I am, in essence, a jerk?

Some of these questions deal with *us*, the other people in your life, particularly those who matter to you. Why does my ex-spouse do those things he does? Can I trust the new associates in my firm? Why was my grandmother so much happier than my mom? Should I be concerned that my daughter invests more in her online "friends" than in her immediate family?

To answer these kinds of questions we will draw on recent advances in the field of personality psychology and explore several key ways of understanding personality. We will start by examining your "personal constructs," the cognitive goggles you use to understand yourself and others. We will then examine your traits, your goals and commitments, and the personal contexts of your everyday life. We will show how each of these factors helps shape the course of our lives and how understanding them helps us reflect on where our lives have gone and where they might still go.

Personality psychology emerged as an academic specialty in the 1930s, but its roots extend back to philosophical and medical theories in fourth-century BCE Greece. Influential among these ancient theories were those that emphasized how various bodily humors—air, black bile, blood, and yellow bile—gave rise to four corresponding temperamental types: phlegmatic, melancholic, sanguine, and choleric personalities. Although such views are now thoroughly discredited, for centuries they were the dominant way of thinking about personality. So if your breakfast conversation had taken place during medieval times, the rumor that you were a jerk, ludicrous or not, would likely have been attributed to your surplus of yellow bile—that was your basic nature, and there was little you could do about it. There are echoes today of such a view about personality in theories that emphasize "types" of individuals. You may have already "typed" yourself because of a test you have taken: you think you are an extravert or a Type A, and you're curious about whether such ways of thinking about yourself have any scientific validity. We will deal in some detail with such issues in the chapters that follow, and the answers might surprise you.

Those of you who have taken courses in psychology will be familiar with theories of personality that emphasize unconscious drives and impulses as the root causes of our behavior. The theories of Sigmund Freud and Carl Jung, which were particularly influential in the early twentieth century, still have an influence in clinical psychology and literary fields but have fallen out of favor among academic researchers in personality psychology. If you have come to believe that the most consequential aspects of your personality are unconscious forces, primarily sexual in nature, what follows will certainly challenge you. Although forces of which we are unaware may well drive our behavior, such influences will not be the prime focus of this book; rather, we will explore how your life is more actively shaped by your goals, aspirations, and personal projects—self-defining ventures that provide meaning in your life. Looking at personality in this way provides you with a vantage point from which to reflect upon your life and think about your future. You are not simply a passive pawn manipulated entirely by forces beyond your control, even though you may have your doubts when you wake up and reflect on what an idiot you were last night.

Another way in which you may have come to think about your personality was through the humanistic psychology of Carl Rogers, Abraham Maslow, and others that flourished in the mid-twentieth century. In contrast with theories that emphasized unconscious determinants of personality, the humanistic psychologists emphasized the more active and growth-oriented aspects of human behavior. A generation that believed deeply in the human capacity, both individually and collectively, to shape our own futures enthusiastically endorsed this human potential perspective. Unfortunately, rigorous science did not match much of the rhetoric of the humanistic movement in psychology. Indeed, scientific objectivity was, itself, seen as a barrier to a true understanding of human nature, a view that was particularly prominent in "new age" approaches to understanding ourselves.

Today this more optimistic view of our capacity for meaningful lives is studied by the field of positive psychology, which explores factors that enhance flourishing in individual lives, communities, organizations, and nations.[1] Positive psychology has explicitly committed itself to a scientifically rigorous approach to understanding human well-being and distances itself from some of the more questionable excesses of humanistic psychology. Although *Me, Myself, and Us* is not a positive psychology book per se, it shares with that field a concern about well-being, happiness, and a sense of meaning in our lives, especially how our personalities influence these aspects of the good life. The application of these lessons from the science of personality to our own personal lives will not be in ten easy steps or formulaic algorithms. It will involve the art of well-being—the creation of a distinctive, singular way of reflecting on your life.

I don't assume in this book that you have prior knowledge about personality psychology or of psychology, for that matter. I only assume that you are curious about understanding how personality can shape our lives. But some of you might well have taken a course in psychology and are aware that personality psychology went through a crisis in the 1970s as a result of the publication, in 1968, of a book by Walter Mischel, then at Stanford University. Mischel's book, *Personality and Assessment*, challenged the whole notion of stable traits of personality. He concluded that there was scant evidence for broad, stable traits of personality, concluding that much of our daily conduct was based instead on the situations we confronted and the ways we construed those situations. Some took this as an indictment of the whole field of personality psychology, and a generation of psychology students was then taught that they should look elsewhere to understand the origins of their behavior. Perhaps you were taught this and thus approach the field of personality psychology with caution.

Today things have dramatically changed. The field of personality psychology is exceptionally buoyant and has expanded into a broad-based personality science studying a considerable range of factors, from neurons to narratives, and drawing contributions from fields as disparate as biochemistry, economics, and literary biography. Within this expanded field the study of traits has been revitalized. I will show you how these enduring aspects of personality have major consequences for your health, happiness, and success in life. We will also see that these traits have a neurobiological base that is, in part, determined by genetic factors. But we will not stop there—personality is more complex than the simple acting out of our biological dispositions. I will introduce you to the distinction between fixed traits and what I call "free traits" of personality, such as when an introverted person acts as an over-the-top extravert, and not only at the office karaoke party. Or a deeply disagreeable person is resolutely pleasant for a whole weekend in October. Perhaps you do this yourself.

Why do you act out of character in this way, and what are the consequences for you?

Beyond the revitalization of trait psychology, contemporary personality science has also made advances in four other key areas. First, our understanding of the biological influences on personality, our first natures, has grown enormously in the past decade. The old dichotomy between nature and nurture has given way to a more intricate and intriguing perspective on how we can nurture our natures. Second, our understanding of environmental influences on personality has been transformed. The social, physical, and symbolic contexts of our lives comprise our second natures. These influences, from our iPod playlists to the "personality" of our cities, both reflect and shape our personalities. Third, there has been a sea change in how psychologists have been exploring the links between personality and human motivation. I have coined the term *third natures* to refer to this shift. Third natures arise out of the personal commitments and core projects that we pursue in our daily lives. Under this new perspective genes influence us as do our circumstances, but we are not hostage to them. Our core projects enable us to rise beyond our first two natures. It is in this distinctively human capacity that the subtleties and the intrigue of human personality are most clearly discerned. Fourth, in contrast with the emphasis on pathology in some of the classic theories of personality, the new personality science is equally concerned with positive attributes like creativity, resiliency, and human flourishing, and in this respect it overlaps with the concerns of positive psychology. The science of personality explores those who are both odd and audacious—strange folks and real characters.

Me, Myself, and Us draws on these advances in the study of personality and examines issues that have consequences for how we think about ourselves and others. Are our first impressions of other people's personalities usually fallacious? Are creative individuals essentially maladjusted? Are our characters, as William James put it, set like plaster by the age of thirty? Is a belief that we are in control of our lives an unmitigated good? Are there patterns of personality that differentiate hardy, healthy people and those at risk for coronaries? Do our singular personalities comprise one unified self or a confederacy of selves, and if the latter, which of our mini-me's do we offer up in marriage or mergers? Are some individuals genetically hardwired for happiness? Which is the more viable path toward human flourishing—the pursuit of happiness or the happiness of pursuit?

Me, Myself, and Us explores these questions and provides a new perspective on human natures and the varieties of well-being. It also provides a framework through which we can explore the personal, more intimate implications of the science of personality. Such exploration may clarify some of the stranger aspects of our daily conduct and help you see your very self and other selves as somewhat less perplexing and definitely more intriguing.

chapter one

First Blushes and
Second Thoughts

Every person is in certain respects like all other people,
like some other people, and like no other person.
Adapted from CLYDE KLUCKHOHN and HENRY A. MURRAY,
Personality in Nature, Society and Culture, 1953

Probably a crab would be filled with a sense of personal
outrage if it could hear us class it without ado or apology
as a crustacean, and thus dispose of it. "I am no such
thing, it would say; I am MYSELF, MYSELF alone."
WILLIAM JAMES, *The Varieties of Religious Experience*, 1902

When I say that Professor Lindzey's left shoe is an "introvert,"
everyone looks at his shoe as if it were something the shoe was
responsible for. . . . Don't look at that shoe! Look at me;
I'm the one who is responsible for the statement.
GEORGE KELLY, *Man's Construction of his Alternatives*, 1958

WHO DO YOU THINK YOU ARE? HOW ABOUT YOUR MOTHER,
your partner, or that strange person sitting across from you in
the restaurant? *Why* do you think about yourself and others in the way
you do? Perhaps you have taken a personality test that "types" you,

but you suspect that you and others you care about are more than this. Perhaps you have heard that the situation you're in, more than the type of person you are, determines your behavior and wonder whether this is true. But types seem too simple and situations too bloodless to satisfy your curiosity about personality. You want new ways of thinking about yourself and others.

Let's start by looking in some detail at the way you typically think about yourself and other selves, at what personality psychologists call your "personal constructs." We will find that how you construe others reveals as much about you as it does about them. And we will find that your personal construing has important consequences for your well-being and how you feel and act in your daily life. Your personal constructs serve as both frames and cages.[1] They can provide some predictable paths through life's complexities, but they can also lock you into a rigid way of thinking about yourself and others. It is possible to change our personal constructs, and this gives us hope. But sometimes escaping from them can be difficult. So let's go back to our initial question: Who do you think you are? Let's see what you make of the proposition that, in an important sense, you *are* your personal constructs.

STRANGERS AND SELF: PERSONAL CONSTRUCTS AS FRAMES AND CAGES

Imagine you are sitting in a restaurant observing the people around you. You notice that one of two men at the next table—the younger, spiffily dressed one, sends his steak back for the third time. Based on your observation of this sequence of acts, what is your first-blush impression of him? Which personal constructs do you invoke?

There are three different approaches you could take. First, you might think of him as having a particular personality *trait*—perhaps assertiveness or extraversion or, less charitably, obnoxiousness. Second, by observing his interaction with his older, grayer table partner, you may infer that the steak returner has an "agenda" or aim beyond procuring meat cooked to his liking: he may be engaged in a *personal project* involving his dinner companion. Perhaps that project is "impress the boss" or "show I don't settle for less than I deserve." Third, you might create a *narrative* that explains his actions. The poor guy has been so demanding tonight because he had a major disappointment at work, and he is lashing out at a server who apparently doesn't understand that medium rare means MEDIUM RARE. You might even use all three approaches more or less simultaneously: the guy at the next table is a demanding jerk who is showing off but clearly has a beef with somebody. Through this process you could have learned more about yourself—and how you assess personality—than about this still-hungry diner.

If you and the steak holder are strangers, as we are assuming here, then attributions about his traits, his projects, or his narrative are suspect. One of the well-documented findings in the study of attributions is that we are more likely to ascribe traits to others, whereas we explain our own actions according to the situations we are in.[2] You have only seen him in this one situation. He might have been acting uncharacteristically, so ascribing a stable trait to him like obnoxiousness could well be unfair. And you certainly have no reliable information on which to assess accurately whether he is trying to impress his boss or whether—and why—he might be feeling bruised and overly sensitive. These are first-blush attempts to explain someone who caught your attention. They are hypothetical hunches based on your personal constructs.

Such hunches are ubiquitous. As Stanley Milgram has observed, in our everyday lives we often make inferences and construct narratives about strangers on the basis of very little information.[3] For example, most of us come in regular contact with "familiar strangers"—people we see in the elevator each morning or in the grocery store or dropping the kids off at school. Our "relationship" with such people is a subtle one. We are aware of each other's presence, but we collude to remain strangers. It is a frozen relationship. And we sometimes create quite elaborate stories about such strangers: he's the guy who looks harried each morning and is probably a divorced lawyer who is ticked off because the Giants lost yesterday. She is a lovely, thoughtful woman who wants to live in Paris, but because she cares for her dying sister, she has foreclosed on her own happiness. And, of course, while you are spinning narratives about them, they are creating stories about you too—about your personality and well-being.

What is particularly intriguing about these frozen relationships is how intensely we resist thawing them, particularly if they have been on ice for a long period. Ask yourself, for example, whether you are more likely to approach a familiar stranger or a complete stranger for the correct time. Unless we meet the familiar stranger in a totally different setting, we are likely to approach a real stranger. But occasionally thawing happens, and we then have a chance to confirm or disconfirm the hypotheses we have been making about our familiar strangers. Sometimes our hunches are bang on, and we take pleasure in having them confirmed. Sometimes our inferences are way off. He's a Green Bay fan, not a Giants fan, and he's happily married, just exhausted by the sleeping patterns of the new twins. The lovely woman isn't really that lovely or thoughtful, and she has been dreaming of living in Peoria and doesn't have a sister. And, again, while you have been creating and revising your construal of these people, they have been doing the same with you. You each have been imputing traits, inferring projects, and weaving narratives.[4] Each of these different ways of making assessments about others—through traits, projects, and narratives—helps us understand personality and well-being. But beyond helping us understand others, they also help us understand ourselves.

The way you construe others has consequences for your well-being. Generally speaking, the more numerous the lenses or frames through which you can make sense of the world, the more adaptive it is. Having too few constructs or insufficiently validated ones can create problems, particularly when life is moving quickly and you are trying to make sense of it. Your constructs can cage you in, and then life does not go as well as it might otherwise.

PERSONAL CONSTRUING AND DEGREES OF FREEDOM

The reason personal constructs matter is because they determine, in part, the degrees of freedom we have for shaping our lives. To explore this in more detail it is helpful to understand the view of human nature that is implied when we look at personality in terms of your personal constructs and then provide some more details about the way in which they influence how we feel and what we do in our lives.

Personal constructs were the key concept in an original and insightful theory of personality written by George Kelly in the middle of the last century. In his two-volume work, *The Psychology of Personal Constructs*, Kelly challenged what were then the two most influential theories about human personality—psychoanalysis and behaviorism. Freudian psychoanalytic theory saw personality as being shaped by the protracted conflict between unconscious, primarily sexual needs and the prohibitions of society. Skinner and the behaviorists believed that what we think of as personality was simply behavior shaped by environmental contingencies of rewards and punishments. Kelly regarded each of these views as providing a far too passive view of the human condition. He proposed instead that each person is like a scientist, actively testing, confirming, and revising hypotheses about people, objects, and events in their lives.[5]

From this perspective, when we form impressions of others we are anticipating how those people will act. The *labels* we use to communicate about our constructs are typically contrasting adjectives. We use them to describe not only ourselves but also our loved ones, professional colleagues, strangers, and the objects we confront in our daily lives. Here are three bipolar personal constructs that you might have used in your daily attempts to make sense of your world: "good-bad," "introverted-extraverted," and "has a USB port–doesn't have a USB port." Clearly, constructs like "good-bad" apply to a vast range of potential objects and events, including cholesterol, body odor, sirloin steaks, and presidential candidates. We say that they have a broad "range of convenience." "Has a USB port" has applicability to a far narrower range of objects, notably electronic devices, and is less convenient to use for construing grandmothers or oysters unless you are dangerously deep into metaphor. The construct "introvert-extravert" is somewhere in between the others in terms of its range of applicability. It is very frequently used in construing people, and its range extends to other creatures such as the neighbor's Maltese terrier. But if someone calls a professor's left shoe "an introvert," it would be more instructive to look at the construer than the footwear to determine what is going on, as Kelly's epigram at the beginning of this chapter so nicely illustrates.

But Kelly also has another point that demonstrates why our personal constructs are central to understanding ourselves. When we construe another person, *we* create the attribute that we then regard as having emanated from the person we are construing. Our inclination to choose particular sets of constructs that we then apply to others can pose problems when they turn out to be inaccurate or simply different from others' constructs. The man in the restaurant did not arrive prelabeled with a tag that said "jerk" or "obnoxious." That was a personal construct the perceiver invoked. Someone else might well have seen the same man and his steak-returning behavior as "classy" or "masculine." In short, our impressions of others' personalities are routed through our *personal* constructs, and these are dynamic, complex, and potentially revisable. Although we might believe that our impressions of others are cool, rational readings of the objects of our construal, personal constructs are frequently hot, emotional expressions of something far deeper.

Consider how personal constructs can influence emotional reactions.[6] Anxiety can be seen as the awareness that something—an event or occurrence, for example—is outside the range of convenience of your personal constructs. If you hear a strange sound at night and it doesn't fit into your typical constructs like "the cat" or "the husband," a blip of anxiety will occur until you are able to confirm another hypothesis, "It's the raccoons again," at which time the anxiety dips. If, however, you sense that it is a burglar, your anxiety will translate into fear, a related but differentiable emotion.

Anxiety may be a more prolonged state, particularly when we are experiencing an unexpected change in our environment, such as the death of a partner, for instance. In this case, life simply can't be navigated in the same old way. New constructs are necessary to make sense of living on your own, to manage changing finances, and to decide whether to keep the subscription to all those sports channels. Who are you now? Those who have more constructs available for anticipating events or the challenges of changed environments are less at risk for experiencing anxiety. Those with very few personal constructs, particularly if those constructs have a very narrow range of convenience, may frequently be upended in their anticipation of events: their constructs just don't apply to many of the new situations they need to deal with in life. In other words, the more limited one's repertoire of personal constructs, the greater the anxiety and the fewer the degrees of freedom one has in anticipating and acting upon events in your daily life. This helps explain why your sister can't seem to move beyond her divorce, in spite of all your attempts to give her new things to do. She treats everyone in terms of a simple construct. "trustworthy vs. will leave me in a flash like Sam did," and in so doing she reduces her degrees of freedom and retreats from re-engaging with life and moving ahead.

Hostility, from a personal construct perspective, is the attempt to extort validation for a personal construct you already suspect has been disconfirmed.[7] Consider a personal construct that you apply to yourself—you see yourself as "dignified" in contrast to those who, in your eyes, "are pushovers." You confront a situation in which you are treated as though you were, indeed, a pushover. In such a situation you may behave so as to force compliance with your own way of construing yourself: You won't back down. You need self-validation. You send back the steak for a second time. And you'll send it back a third time, if necessary, because what is at stake now is not really the steak.

Threat is the awareness of an imminent change in one's core personal constructs. The notion of "core" is crucial here and will figure importantly in subsequent chapters. Personal constructs typically do not bounce around as isolated blips of meaning; rather, they form *systems* with properties that have a profound effect on the way we interpret and act upon events. An important systemic property of personal construct systems is the degree of connection or linkage between each of the constructs in the system. Some personal constructs are relatively peripheral—their use and validation operates independently of other constructs. Others are *core* constructs in the sense that they have strong interconnections with other constructs in the system. They form the foundation of the personal construct system.

Consider how looking at the personal construct system might explain a common experience of parents whose children have gone off to college. A core construct for many students in the first year of college is "intelligent–not intelligent," a construct that can be applied both to themselves and to their actual and potential friends. For some students this construct may be tightly linked to other constructs such as "successful-unsuccessful," "good job prospects–stuck in dead-end jobs," and even "worthwhile-useless." It is possible to assess what researchers call the implicative links between constructs in a system to tease out which are core constructs—those with the greatest implications for other constructs—from those that are more peripheral. Let's assume that "intelligence" is one such highly linked, richly implicative core construct. Consider what happens if an event, such as getting a failing grade on an academic examination, challenges that construct. To the extent that this information disconfirms a person's *core* construct of being intelligent, it is likely to be threatening indeed because it isn't just a single invalidation but rather a challenge to the whole construct system through which that person is navigating life. For a person whose construct of "intelligent–not intelligent" is only loosely linked to other constructs, a failing grade, though disappointing and unpleasant, would not be particularly threatening. Your child is less likely to be devastated by a failed midterm if "achieving well on exams" were not such a core aspect of her construct system. She might learn that being creative and insightful are also worthy features of academic life and incorporate these as core personal constructs.

The emotional consequences of testing and revising personal constructs help us understand how strongly we might resist changing them. The more implications a construct has for other constructs, the more resistance there is to changing it.[8] A few years ago I tried out this idea of understanding our self-conceptions through personal constructs with my students at Harvard. What I learned was how intelligence, at least at Harvard, is linked to construing oneself as "sexy." The class had completed filling out an assessment of their personal constructs and had rated themselves and other people in their social network on each of their constructs. I had told them about a particularly interesting way of looking at resistance to change in their constructs by imagining what would happen to their own self-construal if they woke up tomorrow to find that they were now switched from one end to the other of each of their constructs. You might try this as well: pick the single-most important construct you use to define who you are (e.g., "a good parent," "a New Yorker," "creative"), and now imagine that the opposite defines you. For the students I suggested, by way of an almost random example, that they look at how they would feel if they were no longer at Harvard—that they had never been there. How would that influence their status on other constructs like intelligence, attractiveness, and so forth? The result was intriguing. One of the guys in the class told us that not being at Harvard would have a direct negative impact on his being construed as "sexy." Another student, also a male, agreed, and then another. All males. They all thought they would lose their attractiveness and mate worthiness if they were no longer dressed in crimson. The women in the class looked puzzled and then amused. For two of them the change of status to no longer being at Harvard would *increase* their attractiveness! Regardless of whether it was true, this thought experiment demonstrates both the power and the subtlety of the dynamics of personal constructs and, perhaps, the invidiousness of gendered identities. As much as it might be difficult for a young woman to think of herself as sexy at Harvard, as one of them said on her way out, "at least it isn't MIT."

Gerald: The Man with a Single Construct

Gerald was a student in my class during the early 1970s—a time of peace signs, love-ins, flower power, and an acrid smell of something in the air.[9] On the first day of lectures and every day afterward Gerald stood out from the others. In contrast with the rest of the students' long hair, jeans, and sandals, Gerald (never Gerry) wore a cadet military uniform. He was a blond, husky man who didn't walk into class but, almost literally, marched in. He was seemingly oblivious to some of the other students' looks and snickers. He sat bolt upright throughout my lectures, taking copious notes—a big, erect man at a very small desk. One day I spent the class showing students how to assess the personal constructs they used to construe other people and themselves.

Typically students enjoy this opportunity to explore their personal constructs, and this class was no exception. The most demanding part of the exercise was to calculate each personal construct's interconnections and resistance to change. I walked around the class helping them with the calculations, some of which were quite complicated.

Most students had roughly seven personal constructs that were moderately linked and were, on average, more open to change than resistant. Typical personal constructs the students used in construing themselves and others were "bright–not bright," "interesting-boring," "cool–not cool," "nice-unpleasant" and, in two cases, "groovy-uptight." When I got to Gerald he looked pleased with his analysis and showed it to me. Instead of the usual seven, he essentially had *one* core construct to which every other construct was subordinated—"in the army–not in the army." He applied this construct to relatives, strangers, friends, and, of course, himself. His resistance to changing his own status on the construct was at the highest possible point on the scale. A personal construct approach to personality assumes that, in vitally important ways, "you are your constructs," and for Gerald this seemed clearly to be the case. His internal construing and his external conduct marched in lockstep. He was, then and forever, an army man. This was his very core.

One day later in the term Gerald missed class. He had been so strikingly present in class up to that point that I certainly noticed his absence, though at that point I wasn't particularly worried. However, when he was absent for two more classes, including missing an exam, I became concerned. I found out that he had suddenly dropped out of university and had been hospitalized. Apparently he had been discharged from his officer training program for some disciplinary reason and had, within a few days, ended up in the psychiatric ward, being treated for acute anxiety disorder. Although he may have had other frailties and dispositions that might have made him psychologically vulnerable, from a personal construct perspective there is another compelling explanation: his core self-construct had been invalidated, causing his system as a whole to collapse. If he had been able to invoke other constructs—perhaps a "committed student," "hardworking," "a devoted son"—that would give him an alternative way of seeing himself and his value in the world, then invalidation of his only core construct of being in the army would not have been so deeply unsettling for him. But he didn't, and he fell apart.

HOW DO YOU KNOW? CONSTRUING PERSONS, THINGS, AND SELF

If Kelly is right that we are all scientists, erecting, testing, and revising hypotheses about ourselves and other people, what kind of evidence do we use to carry out the construing process? And what about real personality scientists, not just metaphorical ones—what kind of data do they use? From a personal-construct perspective there is no bright line between everyday lay scientists and "real" ones with PhDs, who actually get paid for testing and revising theories of personality. Of course lay scientists do not generally have access to refined psychological tests or fMRIs (functional magnetic resonance imaging) to use in understanding others' personalities nor do they strive rigorously to obtain consensus about the inferences they are drawing. But there are important overlaps in the data to which they attend when getting to know others. Let's explore how our basic orientations to the world influence the ways in which we assess personality and well-being.

Remember the restaurant and the steak? Ask yourself whether you would have been noticing what was going on at that particular table. Would it have caught your interest? Do you find yourself spontaneously orienting to people of all sorts, curious about what they are talking about, intrigued by their appearance and actions, and wondering about their motivations for what they are doing? If so, you are what I have called a *person specialist*, a kind of George Kelly scientist, but one who has specialized in *other people* as the domain in which you are most engaged. But there are individuals who have a very different orientation, a different specialty. I call them *thing specialists*. They may be looking at the other table, just like you, but are actually focused on the *table itself*, not the people at it. They may be wondering whether its spindly legs will support the massive trayloads of food coming from the kitchen. Or they are intrigued by the new color scheme at the restaurant or the plumbing fixtures in the third cubicle. In short, person specialists are fascinated by people and the world of social relationships. They adopt a personalistic style of knowing others. Thing specialists are intrigued by objects and the world of physical relations. They adopt a physicalistic way of construing the world, including the world of other people.[10]

Whether we are person specialists or thing specialists has implications for how we assess each other's personalities. And this holds both for lay and certified scientists. Those who are person specialists tend to look at others psychologically, in terms of their intentions and motivations. Because these are difficult, if not impossible, to discern without actually talking with people, the person specialist is more likely to engage others in conversation. But if this is not possible, because of practical reasons or the more subtle constraints of being familiar strangers, person specialists are still likely to make inferences about others. Under such circumstances of insufficient information they make unwarranted inferences and may totally misconstrue the other person. Conversely, thing specialists tend to stick with the objective data and are not inclined to infer more than meets the eye. But they misconstrue others by sticking resolutely to that which is immediately apparent, often missing the deeper significance of what they only partially see.

This distinction between personalistic and physicalistic ways of knowing applies equally to the "professional" personality research-ers. Some adopt physicalistic measurement such as fMRIs, physiolog-ical recordings, and genetic techniques to assess personality. Other researchers use more personalistic approaches such as assessing per-sonal constructs, personal projects, and life narratives. These specialist groups seldom talk to each other and can actually get quite grumpy, confrontational, and defensive when presented with the kind of data the other group gathers.[11] Sometimes, such as in executive recruiting, it is desirable and even necessary to get a good take on personality from a variety of perspectives, using many different starting assumptions and specialized measurement techniques.

One effective way of doing this in high-level executive recruit-ment is through the use of *assessment centers*.[12] These are sessions stretching over several days during which candidates go through a diversity of interviews as well as individual and group exercises and social events, all led by a group of assessors (including both experts in personality assessment and senior representatives of the recruiting company). One that I participated in as a consultant was particularly intriguing, as it demonstrated how personal constructs play a key role in how we make decisions.

Derek: Assessing the Tree Whisperer

The client was a giant forest products company that was searching for a senior resource ecologist to join the senior management team. Six candidates had been shortlisted, and the company was holding an assessment center to evaluate them. The position was a big deal. The company was trailblazing a major change in their logging division, with a strong emphasis on sustainable development and what they called ecological awareness, which was a novel idea at the time. The successful candidate was expected to lead this new initiative and, crucially, be able to stand up to the conservative and crusty logging executives who controlled much of the corporate power. But he would also have to stand up to the influential and increasingly radicalized community of antilogging activists. This was an important and highly visible position, and its mandate was daunting.

The assessors set up in the main interview room and prepared for the candidates to arrive. I was seated next to the crustiest of the crusty—Jack Bancroft, a barrel-chested, ham-fisted, fiery-eyed executive who had come up through the ranks and had a considerable reputation for being brusque and blunt. He had once intimidated a consultant who had presented a plan to decrease the acrid smells a pulp mill created by transmitting effluent through underwater channels. "I'm rejecting your proposal because you are obviously a man who has never farted in the bathtub," said Jack as he dismissed the flustered and somewhat confused presenter. So I looked forward with interest and a certain degree of trepidation to what might ensue when the six candidates arrived.

It didn't take long for the sparks to fly. The candidates, all men, arrived and took their seats, and one clearly stood out from the rest. His name was Derek. He was pale and skinny with long, flowing hair, a wispy reddish beard, and watery blue eyes. He had been born and raised until the age of seven in Ireland, when his family had moved to Canada. Unlike the other candidates, he didn't wear a suit; he wore something that resembled a moss-covered smock. Although he didn't wear sandals, he looked as if he really wanted to. Jack's first-blush response was predictable: he audibly snorted, turned to me, and declared, "there is no fuckin' way that hippie is getting this job." I suggested that he should park his expectations at the door and see what unfolded. He gave me a look of withering contempt. For Jack, first impressions meant final conclusions. This was going to be a challenging three days.

Assessment centers are highly intense affairs, and there is little opportunity during the day for the assessors to reflect on all that has taken place. But because after each exercise or interview we had to record our impressions and evaluate each candidate on several criteria, it was possible in the evening to look back at the recordings for the day and see what kind of patterns were emerging. I was particularly interested in Jack's appraisals of Derek. Had that first-blush denunciation of Derek endured? The first day's results confirmed that it had indeed. On each of the first three exercises Jack rated Derek at the very bottom of the group. He was lowest on communication skills, on decision-making ability, and on technical knowledge. In Jack's eyes the only criterion on which Derek shone was creativity. However, creativity was not something Jack valued; it was not an important personal construct. On the general comment sheet that accompanied the report for each candidate Jack had put this comment: "Dick [*sic*] is creative for sure. He talked about being a tree planter and developing a relationship with trees as a young boy. Good Lord! He'll be eaten alive if he gets this position. He is a flake. Mr. Lorax should speak for the trees somewhere else." Jack was not without a sense of humor.

It is true that when the candidates had been asked to say a few words about their motivation for the job, Derek had talked about his childhood delight with walks in woods and other sylvan pleasures. But he had done it in a lightly humorous, almost self-mocking way, as though he were aware of the fact that some in the room really adored clear-cutting and saw trees primarily as lumber. Derek Lorax was no fool.

On the second morning the results from the ability and personality testing were discussed, without the candidates in the room, and Derek's profile was distinctive. Predictably, he achieved very high scores on verbal comprehension and on a measure of cognitive flexibility, the ability to see familiar objects in new ways. And, as expected, he scored at the top of the scale on a creativity test. But he also scored well on analytic skills, particularly in visual processing ability. At the coffee break following the testing feedback I asked Jack how things were going. Assessment centers are designed to provide for independent assessments of candidates, and these are then pooled in the final session, so discussion about candidates before that final session is discouraged. But Jack insisted on telling me that one of the candidates—specifically, Derek—was bugging him. This concerned me. With each exercise Derek had been showing a technical capacity and communication skills that had impressed all the assessors, except Jack. Not only had Jack not shifted in his evaluation of Mr. Lorax; he had actually hardened and polarized his appraisal.

On the second afternoon the group was engaged in a role-playing exercise that proved to be a turning point in the whole assessment procedure. Candidates were to imagine they were at a town hall meeting in which they were representing the company in a debate about forestry practice. The assessors played the role of members of the public who were to pepper the candidates with challenging questions. I played the role of a highly impassioned, partially incoherent, but definitely loud heckler who attacked all of the "spokesmen" with great zeal. Jack joined in, and together we were pretty obnoxious. What then happened took us all by surprise. Derek took us on. He delivered a stirring defense of logging practice, questioned our knowledge of sustainable development, and raised some technical issues that clearly undercut our view of clear-cutting. He was superb. Jack was subdued and went outside for a smoke.

The next session involved having two assessors meet with a single candidate, during which a more intensive examination was given of their motivation for the job and their own personal concerns and stories. I had just recently been developing a way of doing this with a psychological assessment tool I called Personal Projects Analysis and had been encouraged to try it out with the candidates.[13] The technique had a family resemblance to methods used by George Kelly for the assessment of personal constructs, but it focused instead on what people were doing, what personal projects they were pursuing in their lives. If cognitive theorists were concerned with what you're thinking and behaviorists with what you're doing, my approach was essentially asking, "What do you think you're doing?"[14] Jack and I were paired with Derek, and as he began to discuss each of his personal projects he seemed to come alive. He had many projects about which he was enthused, and the content of them ranged from taking a course on financial systems to practicing his bluegrass guitar. It was becoming apparent that although he had some hippie-like characteristics, Derek couldn't be conveniently slotted into a stereotype. In some respects his concerns and commitments were closer to those of the business community, and he expressed an interest, quite rare at the time, in starting his own entrepreneurial venture someday. Whereas in previous encounters Jack had actually turned away from looking at Derek, now he asked him questions and gave flickering signs of interest. He was cautiously gathering data and, I had suspected, having some second thoughts. But then, when he completed his global ratings, he still placed Derek dead last.

The next morning was the final review session. Each candidate was reviewed in turn, and we collated and discussed all the different pieces of information gathered over the course of the past couple of days. We were about to proceed to our final appraisal and global ratings of the candidates when Jack stood up to speak. I had expected an "Anyone but Mr. Lorax" speech, but what we got was deeply different. "I was wrong," said Jack. He then proceeded to say how he originally had been down on Derek, but now he thought he was clearly the best candidate. In fact he thought Derek was a superstar and delivered a passionate plea to rank Derek number one.

I must admit my first reaction was absolute delight. I had seen a man with a well-earned reputation for bombast and first-blush stereotyping have second thoughts. When he looked directly at me and said he had learned a lot over the last couple of days, I was really touched. One of our key goals in assessment centers was to provide a developmental experience for the assessors as well as important feedback to the candidates, and Jack had clearly experienced change. But it was not change I entirely believed in.

Derek didn't get the job. He came in second to a less creative, more reticent, but highly qualified biologist who had impressed the assessors with his blend of technical expertise and soundness of judgment. The candidates didn't get feedback that day, so as we went out to the lobby for final drinks and farewells, there was a general air of anxious joviality wafting about. I joined a small group where Derek and Jack were engaged in intense conversation about bluegrass guitar music. Something intrigued me about Jack and his change of perspective: Had there really been change? It hadn't been a gradual shift; it had been extremely sudden, almost like a tipping point, and then Derek was no longer last but was first. He wasn't a disposable hippie; he was a potential hero of the company. What was going on? Here's what I think now.

Jack in the Box: Slot-Change

Jack, much like our army cadet, Gerald, seemed to have one particularly dominant personal construct, one that was highly evaluative and that was core to him and, in his case, clearly linked to "hippie–not hippie." I suspect that if we had explored this construct with Jack, we would have found that it was linked to a variety of other constructs such as trustworthiness, toughness, reliability, and perhaps even cleanliness. If this is a core construct at the heart of a highly interconnected network of related constructs, then we know that it is resistant to change. We know also that if, for whatever reason, the construct becomes unpredictable or unstable, there can be strong emotional consequences. Before entering the assessment center Jack was very comfortable with the construct of "hippie–not hippie." It had served him in good stead. I had heard that his son had had drug problems and had adopted an alternative lifestyle that had hurt and angered Jack. I knew that Jack had experienced numerous run-ins with environmentalist groups that his division had not handled well, and this had almost cost him his job. I also suspect that Jack was a thing specialist. He was a specialist in machinery. Though not well educated, he was naturally drawn to the more technical side of forestry management. I am pretty sure he didn't read Proust. One thing we know about thing specialists is that they tend to construe not only objects but also people in terms of physical features.[15] Relative to person specialists, they attend to outward appearances, and those images serve as a guide to evaluative judgments.

The guy wears a smock and has long hair? He's a hippie. Period. With all that this entails.

But during the course of the assessment exercises something happens. Jack is exposed to information that is difficult to incorporate into the construct of "hippie." Derek is excited about his project of taking more finance courses, he stands up to environmentalists, he can figure out mechanical things. Good Lord—he's just like me!

There is one other feature of personal constructs that helps explain what happened with Jack's construct system. We call it *slot-change*. If you have a construct system that is primarily centered around one core construct, this means that you have very little wriggle room when that construct is challenged. In other words, to the extent that you have an overly dominant core construct along which much of your construing is organized, you have limited degrees of freedom in navigating your world. If, however, you have many independent personal constructs or several pairs of goggles through which to anticipate events, when one pair doesn't work so well or is invalidated, you can switch to a different construct.

But if you only have one core construct, this means you have only one channel or slot along which you can move when it is upended and only one direction to go: you change ends and, over time, rattle back and forth along this same bipolar construct. If you see yourself, for example, strictly along the construct of "intelligent-stupid" and you fail at something, then the only place for you to go is to the "stupid" end of the construct. And like a car stuck on a snowy road, the further one slides back and forth, the deeper the groove gets and the less likely it is that a new construct—a more adaptive one—might be tried out to anticipate events.

So in one sense Jack had indeed changed in his construal of Derek. But if it were merely slot-change, I worried about what might have happened had Derek been offered the job and they had to work together on important projects. The first time there was an unexpected shift in Derek's behavior there was a good chance he would be slot-rattled back to hippiedom in Jack's eyes. In an important way Jack was boxed in by a core construct that governed at least some of his key relationships both at work and at home.

SECOND THOUGHTS AND DEGREES OF FREEDOM

What we have been discussing throughout this chapter is the value of second thoughts and the importance of having sufficient degrees of freedom in our comprehension of creatures—including ourselves— that we don't slot them away, like William James's crab, as mere crustaceans or, like the above examples, as simply soldiers or typical hippies. Nor ought we see ourselves as *merely* bright or stupid or David's wife or a cat lady. We are free to reconstrue others and ourselves. When we assess individuals' personalities and their well-being we need to take into account information about not only how they are like some other people but also how they are like no other person.

Throughout the following chapters I introduce you to ways of increasing your degrees of freedom to understand yourself and others. We start with looking at people in terms of their relatively fixed stable traits, and we'll see how these have consequences for a person's accomplishments and well-being. But we will also look at what I call *free traits*—the ways we act out of character to advance our core projects. Understanding another person's free traits and personal projects can't be done by standing back and looking at that person dispassionately, like we did with the guy in the restaurant. Instead, we need to engage with the people we wish to know. We don't need a formal assessment center to accomplish this, but we do need to move beyond mere observation and first-blush inference to genuine inquiry and second thoughts.

Is your own way of thinking about yourself centered upon too limited an array of personal constructs? Are there certain constructs that you cling to zealously? Are you threatened when they are challenged? Are you hostile in attempting to validate them? These ways of construing yourself may well be justified and may give you a frame of reference for understanding yourself, but the frame may also limit your capacity for adaptive movement and change when life's situations require it.[16] As you reflect on your own personality and the life you wish to lead, you may also need to explore new ways of seeing and making sense of the other individuals with whom you share your life—your family, friends, and work colleagues. It may be helpful to abandon old constructs, especially for those whom you find confusing or perplexing.

In the chapters that follow I intend to give your personal constructs about yourself and others a good shaking up by having you consider new ways of thinking about personality and well-being. In so doing I hope you experience the satisfaction that comes from increasing your degrees of freedom. In reflecting on your own life and the way you are like all other people, some other people, and like no other person, I want you to be both shaken and stirred.

chapter two

Stable Traits and Well-Being: Set Like Plaster?

It is well for the world that in most of us, by the age of thirty, the character has set like plaster, and will never soften again.

WILLIAM JAMES, *Principles of Psychology*, 1890

It may be that trying to be happier is as futile as trying to be taller and therefore is counterproductive.

DAVID LYKKEN AND AUKE TELLEGEN,
"Happiness Is a Stochastic Phenomenon," 1996

HAVE YOU EVER PUZZLED OVER WHETHER YOU MIGHT BE extraverted[1] or too agreeable or a bit neurotic? Did you readily construe that guy in the restaurant as being overbearing? Do you think your cat is obtuse? If so, you have followed a time-honored way of thinking about yourself and others: you have been adopting *traits* to explain behavior. This way of thinking about people has ancient origins and remains extremely popular today.[2] By invoking traits we are assuming that people have relatively enduring ways of thinking, feeling, and acting that differentiate them from others. In this chapter we'll examine what psychologists have to say about personality traits, with a particular emphasis on what we know about the link between traits and well-being. If the quotations from the two epigrams above are to be believed, both our traits and our well-being are pretty much set in early adulthood and are difficult to budge. So what are these enduring traits, what are their implications for how your life is going, and are they truly set like plaster? I'll begin by telling you about an event at which I was going to lecture on personality traits but was sidetracked by the very embodiment of traits in action before I even began.

PERSONALITY AND PIZZA: IN THIRTY MINUTES OR LESS

I was about to give a talk to a large group of high-tech executives at a Sonoma Desert retreat in Arizona and was on the stage setting up my presentation. A tall, flushed woman bounded up to introduce herself as a member of the planning committee and to make sure, as she put it, that I wasn't "screwing up the AV system." Her curlicued and smiley-faced nametag declared her to be "Deb." Deb was wearing a white T-shirt that had emblazoned on it in big bold red letters *ESFJ*. If you have worked in any moderate-sized organization over the last four decades you probably know what those letters stand for: *E*xtraverted, *S*ensing, *F*eeling, *J*udging—a summary profile generated by the Myers-Briggs Type Indicator (MBTI). Developed by the mother-and-daughter team of Katharine Cook Briggs and Isabel Briggs Myers, the MBTI is a personality assessment instrument based on the theories of Carl Gustav Jung, one of the most influential of the early-twentieth-century psychiatrists.[3] The current version of the standard MBTI is a set of ninety-three questions that assess preferences or tendencies on four major dichotomies: extraversion vs. introversion, sensing vs. intuiting, thinking vs. feeling, and perceiving vs. judging.[4]

The MBTI is enormously popular. More than 2.5 million people are estimated to take it every year. There is a very good chance you have been one of them. It has spawned a flourishing industry selling assessment services, training programs, books, DVDs, T-shirts, mugs, and perhaps even edible underwear, all embossed with the four-letter profiles. Why is the MBTI so extraordinarily popular? And why did I inwardly roll my eyes when I saw Deb's T-shirt? Is the MBTI popular because of its reliability and validity? Probably not. The four-letter code, like that on Deb's T-shirt, represents one of sixteen "types" based on binary scores on the four preferences. The problem is that the consistency with which you are likely to receive the same profile score on repeated assessments is weak.[5] In other words, the Myers-Briggs lacks *reliability*, and it is likely that your particular four-letter profile will shift from one testing to the next.[6] So Deb may want to invest in some additional T-shirts. With respect to *validity*—whether the test is measuring what it purports to measure—the MBTI is adequate but not exceptional and does not have the extensive research base that other personality tests have. So why is it so extraordinarily appealing both to organizations and individuals?

I think there are five reasons. First, it is easy and enjoyable to take the MBTI. Workshops organized around MBTI assessment can be great fun for most participants and an effective way to engage groups in team-building activities. Consider one reviewer's account of organizations' use of the MBTI:

> A corporate trainer in the Atlanta area, who asked not to be identified, worries that her organization has gone "type happy". . . . The vogue in this organization is brownbag lunches and MBTI. "It's a little like a mass horoscope reading or something," she says. In other words, it's quick and easy. "First you call Domino's and then you call the training department. We both deliver in 30 minutes or less."[7]

Such a depiction of psychological assessment makes academic personality researchers wince. In terms of my own personal constructs, the notion of a horoscopic-like device delivered with pizza delivery speed is the polar opposite to the nuanced and detailed analysis I believe to be essential for understanding human personality. But for everyday purposes millions of people find great value in having a four-letter summary of themselves that can be easily assessed, worn on their sleeves, and stuck on their mugs.

A second and related reason for the MBTI's popularity is its marketing and packaging of materials and spin-off products. They are colorful and glossy and create an aura of professionalism (or, to some eyes, marketing savvy) that other personality assessments seldom match. A third reason is that sharing and comparing MBTI profiles opens up possibilities for conversation about personalities and preferences that can generate genuine insight, unlike similar conversations about horoscopes. Fourth, people readily identify with their personality profiles, whether they are presented in the form of MBTI-like profiles or as scores on more finely differentiated dimensional scales. It becomes *part of their identity*, way more than their favorite pizzas are. Deb clearly identified with her appraised personality and wore her MBTI profile with great pride, almost as a badge of honor.

There is a fifth reason for the appeal of this kind of feedback about personality that applies not only to the MBTI but also to other forms of assessment. This is something I call the *magical transformation*, and it occurs when we contrast how individuals typically feel when they are answering questions on personality tests with how they feel when they see their final profile. Perhaps you have experienced this if you have taken such a test. When answering the questions did you find yourself saying, "This is frustrating. It all depends on my mood or the situation?" But after your answers are scored, your profile appears, and you recognize yourself in it, you say, "That really *is* me!" There is something about seeing one's personality captured in a profile that seems to whisk away the skepticism and stimulates immediate interest, even intrigue.[8] And this is what I think happens with the MBTI. Although people may be rather skeptical when completing the items, most really enjoy the output and want to share it with colleagues, family, and friends. And I should also note that with the MBTI there are no "bad" profiles—each is described in terms that can evoke admiration. This is a personality profile you can proudly tell others about, and this too, no doubt, contributes to its popularity.

So why did my eyeballs start tilting upward when I saw Deb's *ESFJ* bouncing around in front of me? Deb, bless her heart, was the kind of person I find difficult to deal with when I am just about to approach the podium for a presentation. It was primarily the letter *E* that alerted me: Deb was an extravert. She was a blunt and blustery person who was taking charge of handling me and my AV requirements in what she thought was a helpful fashion. I tend not to be a blunt and blustery person, I don't really like to be "handled," and my AV was working just fine, thanks. If I were the kind of person who wore his personality profile on his chest— and I'm not—my four-letter code would be the opposite of hers, especially the first letter. "Hi," it would read, "I'm Brian, and I'm an Introvert." But it isn't through the MBTI that I identify as an introvert; it is through research based on one of the most influential of contemporary research areas in personality science—the five-factor model of personality.[9]

THE BIG FIVE TRAITS OF PERSONALITY

Before introducing you to what contemporary research in personality has discovered about personality traits, including introversion/extraversion, I thought you might want to take the following:

TEN-ITEM PERSONALITY INVENTORY (TIPI)

Here are a number of personality traits that may or may not apply to you. Please write a number next to each statement to indicate the extent to which you agree or disagree with that statement. You should rate the extent to which the pair of traits applies to you, even if one characteristic applies more strongly than the other.

Disagree strongly	Disagree moderately	Disagree a little	Neither agree nor disagree	Agree a little	Agree moderately	Agree strongly
1	2	3	4	5	6	7

_____ 1. Extraverted, enthusiastic. 5

_____ 2. Critical, quarrelsome. 4

_____ 3. Dependable, self-disciplined.

_____ 4. Anxious, easily upset.

_____ 5. Open to new experiences, complex.

_____ 6. Reserved, quiet.

_____ 7. Sympathetic, warm.

_____ 8. Disorganized, careless.

_____ 9. Calm, emotionally stable.

_____ 10. Conventional, uncreative.

The TIPI was developed by Sam Gosling, Jason Rentfrow, and William Swann. For source and scoring instructions see Notes.[10]

The scale you have just taken—which, it should be noted, beats the pizza delivery time by about twenty-eight minutes—is used by personality researchers wanting a very brief, but reliable and valid indication of your status on five important and consequential dimensions of personality, conventionally called the Big Five traits. I provide it simply to give you a hint about your own personality as viewed from the trait perspective. I should also point out that I will be presenting different personality scales throughout the following chapters. These scales are meant solely for self-reflection. They have been developed as research tools and, in some cases, for use in teaching courses in personality. They are not intended for use as diagnostic tools and should be interpreted cautiously. For those wishing a longer and more comprehensive assessment of the Big Five, the gold standard is the NEO PI-R, developed by Paul Costa and Robert R. McCrae, and it has been used extensively in studies around the world.[11] Both the longer and shorter versions of the Big Five scales reflect the consensus among personality researchers that the diverse ways we differ in personality can be effectively reduced to five major factors—Conscientiousness, Agreeableness, Neuroticism, Openness, and Extraversion—referred to sometimes with the acronym CANOE.[12] Unlike the MBTI, these terms are not meant to signify different *types* of people in which one is or is not, say, an extravert or a neurotic; rather, the Big Five traits are *dimensions* along which all people can be placed, with most falling in the middle regions and others spread out along the full range of scores.

There is clear evidence that each of these dimensions of personality has a genetic component accounting for roughly 50 percent of the variation between people on each trait.[13] It is also clear that how people score on these dimensions of personality has important consequences for happiness, health, and human achievement—all core components of well-being.[14] This takes us back to the question of whether personality and well-being are, in William James's phrase, "set like plaster." Is this a reasonable way of taking stock of your life to date and your future prospects? With which aspects of well-being are the Big Five traits associated?

In the following sections I will present the case for looking at lives in this way and, in so doing, suggest you try these on as personal constructs. I will examine four of the Big Five traits and describe the diverse aspects of well-being that are associated with them. I will then provide a more detailed analysis of the fifth trait of extraversion in order to explore some of the subtleties of the link between stable personality traits and well-being.

Conscientiousness: Structure, Chaos, and All That Jazz

If you scored high on the scale of conscientiousness, you likely would be described as well organized, orderly, careful, persevering, prudent, circumspect, and nonimpulsive. In contrast, those scoring low would likely be described as disorganized, spontaneous, careless, imprudent, and impulsive. At first blush it would appear that being high in conscientiousness is a very good thing indeed. And sure enough, there is a considerable amount of research evidence that indicates being high on conscientiousness augurs well for many diverse aspects of well-being.[15]

Conscientiousness is strongly related to measures of successful achievement in the academic and occupational domains. For example, student scores on conscientiousness are a better predictor of college grade point averages than are high school grades, which are widely thought to be the best predictors. Conscientiousness also is a strong predictor of retention in college. The reason for this seems pretty straightforward: the everyday ecology of higher education rewards those who can meet deadlines, forego the diverting delights of campus life in order to study for exams, and control the impulse to imbibe too much.[16] And when the time comes to leave college and enter the job market, it is easy to see why those low on conscientiousness are unlikely to impress. Being seen as impulsive, lackadaisical, and imprudent as well as having arrived late for the interview is not going to endear you to most recruiters. Once again conscientiousness predicts a valued outcome, greater likelihood of getting a job, and, once in it, predictably higher job evaluations and larger salaries. What is remarkable is the extensive array of different occupations for which conscientiousness predicts success and achievement. In short, being high on conscientiousness is one of the very best predictors of conventionally defined success and, in that respect, an important determinant of our potential well-being.

Being conscientious is not just associated with greater educational and occupational success; it is also a strong predictor of health and longevity. Consider your score on conscientiousness and imagine how your teachers and parents would have rated you on that scale when you were eleven years old. Howard Friedman and his colleagues provide some intriguing evidence that such ratings predict longevity.[17] Indeed, the positive impact of conscientiousness is the equivalent of the negative effects of cardiovascular disease on life expectancy. Just as there is a link between conscientiousness and achievement at school and work, it appears that conscientious people live longer due to the kinds of health-related tasks and projects they engage in throughout their lives. Those who are high in conscientiousness are more likely to pursue and adhere to everything from flossing to fitness regimens.

It appears, then, that conscientiousness is a clear and consistent predictor of well-being, at least at first blush. But might there be a downside to this particular personality trait? There are some indications that there may well be.

Daniel Nettle has made a convincing case that conscientiousness is adaptive primarily in environments, or social ecologies, that are predictable and well ordered.[18] The ability to persist in projects and tasks that require committed pursuit and timely completion come easily to highly conscientious people. However, if an environment is chaotic, unpredictable, and fast paced, it is possible that conscientiousness could be maladaptive. In such environments the less conscientious person might be better able to orient away from routine activities, to attend to sudden intrusions, and to change direction with alacrity.

Some confirmation of Nettle's view is contained in research by Bob and Joyce Hogan that examined the relation between rated conscientiousness and job effectiveness with a particularly interesting group of professionals.[19] Contrary to the vast bulk of studies, they found that those who were rated as highly conscientious obtained significantly *lower* evaluations of effectiveness. Why the discrepancy? The Hogans had studied Tulsa jazz musicians. For these musicians, lower conscientiousness was associated with being rated as a better musician by their peers. Imagine, for example, what happens in improvisational jazz or among musicians in a club band who may never have played together before. Beyond the traditional repertoire, which each musician is familiar with, there is unlimited scope for creative variation. A highly conscientious jazz musician would likely know the standard repertoire extremely well but could be insensitive to subtle cues from others intimating a change of cadence, rhythm, or key. When playing the national anthem most of us would prefer someone very high on conscientiousness, not someone who is likely to wander off into random notes of whimsy, however creative they might be. But when the environment is fluid, the structure open-ended, and the players and audience attuned to new possibilities, the virtues of unremitting conscientiousness might well be questioned.[20]

Such considerations are not limited to musicians. Consider your own particular social ecology. Is it more like that of a well-organized, formal, hierarchical organization? Or is it more like improvisational jazz? When evaluating the link between personality and well-being we

need to take into account both the individual personality and the social ecology within which that person is acting. A seemingly positive
trait may be adaptive for a limited array of tasks and projects. This has
consequences for the way we take stock of our lives and chart new directions for ourselves.

Agreeableness: The Promise and Problems of Being Pleasant

The second of the Big Five traits is agreeableness. Highly agreeable people are seen by themselves and others as pleasant, cooperative, friendly,
supportive, and empathic. In contrast, their disagreeable counterparts
are seen as cynical, confrontational, unfriendly, and mean-spirited. It is
obvious that agreeableness is regarded as a highly desirable personality
trait, particularly in contexts where the person is working with others.
It is the dimension of the Big Five that people are most attuned to when
they form first impressions of others. Some speculate that scanning for
whether a person is agreeable or disagreeable answers a question with
a long evolutionary tail—can I trust this person to be an ally?[21]

Despite its importance in the way we form impressions of others,
agreeableness is not associated with success in the way that we found
with conscientiousness. Indeed, compared with the other Big Five traits,
agreeableness is one of the weakest predictors of organizational success.
In fact, there is even evidence that agreeable people are *less* successful
in their working life, as indexed by their salaries. This is particularly
the case with men, for whom agreeableness may run counter to norms
of masculine conduct.[22] Once again, however, a social ecological issue
needs to be raised that asks whether there may be some work environments in which this tendency for agreeable people to perform less well
might actually be reversed?

A recent study carried out in Finland provides some new data that
contrasts with the prior evidence about job effectiveness and agreeableness.[23] It found a strong and consistent relation between being agreeable and a diversity of measures of effectiveness. Why the discrepancy
with previous research? The group that was studied happened to be

key account managers, whose job is to nurture relationships with important clients who are critical to the organization's sales and development. Although agreeableness may not be effective in short-term or initial encounters with others, it may have greater effectiveness in cases in which measures of effectiveness are based on long-term relationships rather than initial impressions. In short, the impact of personality on performance needs to take into account the temporal and social ecology of that impact. Timing and context matter.

There is another possible reason why the link between agreeableness and effectiveness is equivocal. It may be that being too agreeable or too disagreeable are *both* associated with poorer performance and that an optimal, middle-level degree of agreeableness exists. In other words, both nice and nasty guys (and their sisters) may finish last, but only if they overdo it. There is some evidence that supports this hypothesis. Disagreeable people, along with extraverts, are both characterized by assertive behavior. Assertiveness involves a trade-off between ease of relationships and goal achievement. Getting ahead may involve difficulties in getting along. When individuals were asked to evaluate leaders' effectiveness, those who were seen as too high or too low on assertiveness were rated as less effective; those at the optimal level were rated most effective. Relatedly, research has found that disagreeable people differ from extraverted ones in that, although they are both assertive, those who are highly disagreeable fail to differentiate between important and unimportant situations and tasks. They are indiscriminately assertive, whereas extraverts make finer discriminations.[24]

The link between health and agreeableness is also complex. Highly agreeable people are more likely to create social networks, which provide an important resource for enhancing health. Individuals who are low on agreeableness not only suffer from not having close social ties to draw upon; they are also at direct risk for health issues relating to their disposition to anger, cynicism, and antagonism.[25]

There is another rather intriguing aspect of disagreeableness that goes to the issue of happiness. Although agreeable people are more likely to report they are happy, disagreeable people are more likely to say they are happy when they are being disagreeable! In one study that involved "beeping" people with a pager at random times during the day, disagreeable people were more likely to express positive emotions when they were engaged in acts such as disciplining others than when they may have found themselves in inexplicably pleasant surroundings.[26] We will explore further aspects of hostile behavior and cardiovascular risk in Chapter 6.

Neuroticism: Sensitivity and Sensibility

Neuroticism vs. stability is one of the most extensively studied personality dimensions and one of the most critical for predicting diverse aspects of well-being. Although we have seen that conscientiousness and agreeableness have complex relations with well-being, the neuroticism-stability dimension is fairly straightforward. Those who score on the neurotic end of the dimension also score low on many different facets of positive functioning: they have lower subjective well-being, more negative than positive emotions, difficulties in marriage and interpersonal relations, less job satisfaction, and compromised physical health.[27]

I should stress that we are not talking about neurotic illness here; we are talking about a dimension along which normal people vary. The central core of neuroticism is sensitivity to negative cues in the environment. There is a clear neurological basis for this sensitivity: neuroticism is associated with hypersensitivity of the amygdala, a structure that alerts organisms to the presence of threat. Those scoring high on the neuroticism scale detect, recall, and ruminate on perceived threats, dangers, and slights that a more stable person would not see.[28] By being ever vigilant to the possibility of threat, real or imagined, neurotic individuals experience high levels of chronic stress, which can challenge the immune system and create a risk for physical illness.[29] They have more sleep disorders and more frequent doctors' visits, and they generally report more health problems. Although it is important for all of us to be alert to the signs of threat or danger in our environments, neurotic individuals are hypersensitive to such cues. Consequently those who score high on the neuroticism scale are more prone to anxiety, depression, self-consciousness, and emotional vulnerability. In contrast, of course, those at the other end of the scale—"stable" individuals—are more robust and less vulnerable to the vicissitudes of everyday lives.

Neuroticism also might play an important role with respect to the other personality dimensions. It can be thought of as an *amplifier* of other dispositions. For example, those who are conscientious and highly neurotic are more conscientious than if they were low on neuroticism and may be prone to obsessive-compulsive type behavior.[30] Those who are disagreeable and highly neurotic are at risk for being deeply, perhaps dangerously antagonistic.

Given the pervasive problems experienced by those who are characterized by a neurotic personality, it is tempting to be pessimistic about their quality of life and well-being. Conversely, stable individuals would seem to have a clear path toward well-being and a flourishing life. But once again we might consider whether there are costs and benefits at *both* ends of the spectrum.

It is interesting to speculate on the evolutionary background of neuroticism.[31] What selection pressures led to neurotic individuals being here at all? I think their sensitivity is pivotal. Although it is true that sensitivity can be debilitating, it also has had a vital adaptive function since the very beginnings of human evolution. The varieties of human personality emerged during the Pleistocene, when our forebears were hunting and gathering and living in groups of about thirty. The conditions were challenging, so having individuals who were particularly sensitive to threat played a valued role. Besides alerting members of their group to possible danger, neurotic individuals were more likely to detect and avoid predators themselves. Their happier stable friends were more likely to become imperiled prey. Although the nature of threat is different now from then, it is still pervasive and the sensitivities of neurotics may continue to exert a protective influence.

Openness to Experience: Receptivity vs. Resistance

The trait of openness versus being closed to experience refers to the tendency to be receptive to new ideas, interactions, and environments and is closely linked to creativity. Those scoring high on openness have artistic and cultural interests, a preference for exotic tastes and smells, and a more complex way of construing the world. In contrast, those who score low on the openness scale are more resistant to trying out new things, are comfortable with routines, and find the lure of the exotic unalluring and the untried rather trying. Openness applies as well to experiencing emotions. Like neurotic individuals, open ones are more likely to acknowledge negative feelings of anxiety, depression, or hostility than are more closed individuals. But in contrast with their neurotic counterparts, they are also more likely to experience positive emotions such as delight, wonderment, and joy.[32]

One particularly interesting example of positive emotions is that of experiencing aesthetic chills, which has a unique and strong link with openness to experience. Do you frequently feel the hairs on your back rising (even if your back is relatively bald) in response to hearing a particular piece of music or viewing a particular work of art? We call those pilo-erections (literally "hair standing up"), and if you experience them frequently, the chances are you are high on the trait of openness to experience.[33] I have to admit to having such pilo-erections particularly with respect to music. Given the heritability of the Big Five dimensions of personality, I was intrigued to discover over the years that my daughter, Hilary, and I shared not only a receptivity to certain kinds of music but even to specific passages within a particular piece of music. Frequently we would send each other music clips and predict when we would each get an aesthetic chill. One spring day Hil had brought over a CD that contained a passage she was sure would create a pilo-erection for me. At precisely the same time as the evocative passage was wafting through the living room my granddaughter entered the room, shuddered, and said, "It's really cold in here." It wasn't cold; it was 70 degrees. I was convinced she had experienced an aesthetic chill. When Hil and I saw what was happening we both got another round of chills—we experienced pilo-erections about experiencing pilo-erections.

With respect to well-being, openness has a rather different set of linkages from the other Big Five factors we have examined so far. As mentioned, openness is associated with both positive and negative emotions, so, on balance, open individuals may have a more nuanced sense of well-being. We will discuss the relation between openness and creative accomplishment in detail in Chapter 7, but for now it can simply be said that a disposition to openness is likely to be associated with success in endeavors and occupations that place a premium on innovative accomplishment.

Extraversion: Arousal and Affect

I want to spend more time on the Big Five dimension of extraversion-introversion for several reasons. It is, along with neuroticism, the most studied of the major dimensions of personality and one of the most consequential for understanding well-being. There has been a great deal of buzz recently about introversion-extraversion, in large part due to the publication of Susan Cain's *Quiet: The Power of Introverts in a World That Can't Stop Talking*.[34] Her central argument is that in the United States in particular there is an Extravert Ideal that creates systematic biases, from kindergarten classrooms to corporate boardrooms, against introverted styles of behavior. The book has struck a resonant chord among readers and it is important that we understand some of the features that distinguish extraverted from introverted personalities. Like other dimensions of the Big Five, extraversion is known to have a moderately high degree of heritability. One biological model of this dimension postulates that differences in extraversion reflect differences in the arousal level of certain neocortical areas in the brain: those high in extraversion have low levels of arousal, whereas introverts have high levels.[35] Given that effective performance on daily tasks requires an optimal level of arousal, extraverts are typically seeking to increase their levels of arousal, whereas introverts are trying to lower theirs.

In everyday interactions introverts may avoid highly stimulating settings because they realize, perhaps only tacitly, that their performance is often compromised in such environments. When observed doing this they may be misconstrued as being antisocial. Conversely, extraverts seek out arousing settings precisely because they have learned that they perform better when engaged in the cut and thrust of animated, even heated exchanges. In my experience these differences are often manifest in people's behavior behind the wheel of their cars, and they are easy to spot. Imagine a car containing one introvert and one extravert driving along the highway. Typically it is the extravert who is driving—even if it is the introvert's car. The extravert drives in such a way as to increase arousal. They drive quickly, often too quickly, and are more prone to accidents (and traffic tickets).[36] Despite local ordinances against it, I suspect that extraverts use cell phones in their cars, perhaps several phones simultaneously, in order to keep alert and awake, while the introverted passenger looks grimly ahead, hoping to arrive at the destination in one piece. Both can do this with relative impunity; it is not a zero-sum game. What *is* a zero-sum game, in which my win entails your loss and vice-versa, is in the negotiation of the radio in the car. Extraverts are inclined to turn the radio up to about 110 decibels, which is near the pain threshold—at least to introverts' ears. And pain is the operative word. Introverts actually have greater sensitivity to pain than do extraverts, particularly if they are also neurotic.[37] When I used to coach kids' soccer (badly) I would tell parents to ease up on criticizing their introverted kids if they complained of an errant kick from an opposing player. I often thought that extraverted kids actually enjoyed getting whacked from time to time—it didn't seem to faze them.

Besides increasing or decreasing the stimulation level of the environment, you can also achieve an optimal level of arousal by drinking beverages that have a direct impact on neocortical arousal.[38] Alcohol, at least initially, has the effect of lowering arousal. After a couple of glasses of wine the extraverts are more likely to dip below the optimal arousal level, whereas their introverted friends, nudged closer to optimal arousal, may appear unexpectedly garrulous. Coffee, being a stimulant, has the opposite effect. After ingesting about two cups of coffee, extraverts carry out tasks more efficiently, whereas introverts perform less well. This deficit is magnified if the task they are engaged in is *quantitative* and if it is done under time pressure. For an introvert, an innocent couple of cups of coffee before a meeting may prove challenging, particularly if the purpose of the meeting is a rapid-fire discussion of budget projections, data analysis, or similar quantitative concerns. In the same meeting an extraverted colleague is likely to benefit from a caffeine kick that creates, in the eyes of the introverts, the illusion of competency.

It should be remembered that this Big Five dimension of personality, unlike the "is-isn't" categorizing of MBTI profiles, is measured on a *continuous* scale, with most individuals obtaining middle-level scores. We often refer to them as ambiverts, and chances are that you are one. In terms of arousal level, ambiverts are chronically close to the optimal level, in between introverts and extraverts. There is also some recent research that shows an "ambivert advantage." The organizational psychologist Adam Grant has found evidence that, contrary to the common assumption that extraverts are the best people in sales fields, ambiverts do better than either extraverts or introverts.[39] I suspect that future research will reveal this advantage in other areas as well. And I do think I owe ambiverts at least a hint as to which kind of beverage they should drink to maintain their tendency to be at an optimal level of neocortical arousal: I suggest Irish coffee. Or water.

Differences in extraversion also play a role in intellectual achievement.[40] Generally speaking and, except for one elementary school grade exception, introverts achieve higher marks in school, so by the time they are in university they are more likely to obtain a first-class graduating average. Why is this? Could it be that extraverts are simply less intelligent? The research suggests this is not so—there are no reliable differences in IQ between those scoring high and low on extraversion. I believe it is the learning environment that is critical. Extraverts learn better in environments that are stimulating and engaging, and conventional schools may not be able to provide such an environment. Consistent with the notion that engagement is central for extraverts, the introvert advantage in marks disappears when we look only at laboratory classes. And the elementary school grade exception in which extraverts come home with a better report card? Kindergarten. Though tempting, it is probably not wise to predict later academic achievement on the basis of how our children did in kindergarten. Our extraverted children may well have peaked then!

There are two other areas of intellectual achievement in which there are notable differences between those who are high and low on extraversion. Extraverts have better memories than introverts do, but only in short-term memory. Introverts do better on long-term memory tasks.[41] Also, when we engage in tasks, we can adopt two different strategies involving a quality-quantity trade-off: we can do things quickly and make a few mistakes, or we can do things slowly and get it perfect. Extraverts are more likely to opt for quantity, introverts for quality. These intellectual and cognitive differences can give rise to conflicts or at least mutual eyeball rolling between colleagues, especially when they are working on joint projects. Introverts, preferring a slow and careful approach to their tasks, see their extraverted colleagues as too "crash, bang, wallop" and want to rein them in. Extraverts can get exasperated at their introverted colleagues' style; they want them to speed up and get things done, even if there are a few little mistakes. When such creatures are housed together periods of protracted pique can ensue.

If we watch social interactions, we can easily spot the difference between introverted and extraverted styles. Their nonverbal interaction styles differ sharply. Extraverts stand closer but speak more loudly. They tend to touch and poke, even hug. Introverts are less intense, more subdued, and definitely less huggy. As a result of these differences, when extraverts and introverts interact, it can look like a rather bizarre dance—a series of alternating lunges, retreats, pokes, and aversions.

They also have contrasting verbal styles. Extraverts use direct, simple, concrete language. Introverts have a tendency to craft communications that are more oblique, contingently complex, and weasel-worded (more or less, at times, or so it appears). Such differences can create all manner of friction between well-meaning friends and family members, with much rolling of eyes and gritting of teeth.

Here is an example. I once had a consulting contract with a colleague who is about as different from me as two folks can be. His name was Tom. He is six foot four, whereas I have a tendency not to be tall; he's an extravert, I'm an introvert. The client had arranged for a man from the finance division, let's call him Michael (because that was his name), to join our project team for a month. His personality and style almost brought the project to a grinding halt. When it became obvious that something was amiss, the client asked Tom and me what we thought of Michael. Tom responded in classic Extravertese (to be revealed in a moment). When asked what *I* thought about him, I paused and then said something like, "Well, Michael has a tendency, at times, to behave in a way that some of us might see as, perhaps, rather more assertive than is normally called for." Tom rolled his eyes and declared, "Brian, that's what I said—he's an asshole." Now, as an introvert, I might gently allude to certain assholic tendencies in Michael's personality—and Tom's, for that matter—but I am not going to lunge for the A-word. We introverts tend to hedge our comments in ways that will protect us from eventual disconfirmation. We tend to be oblique. Extraverts don't really do oblique.

The dimension of extraversion-introversion also helps us understand motivation and the ways in which we scan our environments. Just as neurotics have a sensitivity to punishment cues, extraverts are highly sensitive to reward cues and reward opportunities. When they look out at their environments they see the positive possibilities around them. Reward cues do not motivate introverts as much; indeed, particularly if they are also neurotic, introverts are hypersensitive to punishment cues. Extraverts and introverts can see virtually identical events and construe them in radically different ways.

My favorite example of this was told to me by a pediatrician who had been advising young mothers about food allergies and sensitivities. Two mothers during the same week had reported in for a discussion about the diets of their eighteen-month-old children. Each had mentioned some distinctive features of their children's preferences, and, rather surprisingly, a strong preference for ketchup had been reported for each child. In the consultation the first mother had expressed concern about the problem of the child not being able to eat *anything* without ketchup. She seemed worried that it portended some gastrointestinal abnormality, perhaps early stage Heinz disease, and wanted guidance on how she should handle it. The next day he met with the other mother. So does your child have any eating problems, he asked? "No, not really." "No?" "Hell no, just give the kid some ketchup, and he'll eat anything!"

It should be clear at this point that extraverts are well poised to have happy lives. If we look at measures of positive emotions, life satisfaction, perceived quality of life, and success in fields that place a premium on social engagement, extraverts, with all other things being equal, appear to be flourishing. Even when it comes to sexual behavior it appears that extraverts are at an advantage. In a study of the frequency per month with which intercourse occurred, introverted males reported 3.0 times, extraverted males 5.5, and introverted women 3.1. As for extraverted women, speaking as a male introvert, I think they are heroic: 7.5.[42] They not only handle all the male extraverts, they pick up a few introverts as well! But to assuage any concerns that introverted male readers (or their partners) might have about the results, I ask you to recall the earlier discussion about a *quality*-quantity trade-off.

So what do we make now of the debate Susan Cain's book provoked about whether introverts have been systematically discriminated against in American culture and, to a somewhat lesser extent, in other modern Western countries? In some respects introverts are indeed given short shrift. As Cain describes in compelling detail, many classrooms are designed for group activities that work, as we have seen, to the disadvantage of introverts. Many of the professional schools in business administration and related fields place a premium on the extraverted interaction style: fast, intense, and, well, noisy. As a consequence of a variety of such pressures against introverts, have they been cheated of prospects for a happy life relative to their extraverted friends and peers? I think on balance Cain's conclusions are compelling and dramatic. She calls for a shift in consciousness that would empower introverts just as women were empowered a few decades ago.

However, one issue that this chapter ought to have brought home is that, although extraversion-introversion is an exceptionally important personality trait dimension, it is only one of five major traits that personality researchers have identified. Consider two extraverts who differ on some or all of the other four factors of personality. An extravert who is open, agreeable, and stable is a very different creature from one who is closed, disagreeable, and neurotic. In short, the conversation on the politics of personality needs to consider traits beyond extraversion.

I can now let you in on the dynamics of coping with a card-carrying, self-proclaimed extravert like Deb just before giving a presentation. Although I have had years of experience talking to large groups, as an introvert, I typically need to lower my level of arousal before giving a presentation, typically by taking a walk or just looking over my notes quietly in a room off the main lecture hall. So as soon as Big Deb leapt on stage three minutes before I was "on," my level of neocortical arousal started rising. She was also blunt in a way that most extraverts would actually see as quite delightful but that, for me at the time, seemed rather gratuitous and presumptuous. I'm a conscientious introvert—I don't screw up the AV! So that also caused a blip of neocortical arousal. And she was obviously a Myers-Briggs aficionado, and I am resolutely opposed to putting people in pigeon holes as either introverts or extraverts, traits set like plaster, as William James put it. I am convinced that we have the capacity to adapt our personalities to the demands of the day and to enact our social selves in ways that advance the things we care about. I think William James got it only 50 percent right. I think that we humans are essentially half-plastered. I assumed Deb would have disagreed with me.

I didn't quite have it right, however. As the audience settled into their seats and Deb and I were wrapping up the AV setup, she turned around and softly said, "Scared you, didn't I?" And then in a kind of conspiratorial whisper she told me she had been a student in one of my large classes many years earlier and that she was having a bit of fun with me. As she turned around to walk off the stage I could see clearly, etched on the back of her T-shirt in soft blue print, the four letters *ISTP* (Introverted, Sensing, Thinking, Perceiving), the exact opposite of what was on her front. She knew what I was going to say in the presentation. I was going to convince the audience, as I hope to have convinced you, that personality traits are stable, consequential determinants of our health and happiness and achievements. But then I was going to turn the tables on them and let them in on a secret, one that I will let you in on as well in the next chapter.

How you score on the Big Five dimensions of personality does have consequences for your well-being and achievements in life. Your personality traits have a genetic base and are relatively stable over time. But does this mean hardwired traits constrain your degrees of freedom in shaping your life and that attempting to change is futile? Let's see.

chapter three

Free Traits:
On Acting Out of Character

There are times when I am so unlike myself that I might be taken
for someone else of an entirely opposite character.

JEAN-JACQUES ROUSSEAU, *Confessions*, 1782

And I have known the eyes already, known them all—
The eyes that fix you in a formulated phrase,
And when I am formulated, sprawling on a pin,
When I am pinned and wriggling on the wall,
Then how should I begin
To spit out all the butt-ends of my days and ways?
And how should I presume?

T. S. ELIOT, "The Love Song of J. Alfred Prufrock," 1920

S O FAR I HAVE TRIED TO CONVINCE YOU THAT YOU TOO ARE
a scientist actively construing your world with your own set of
working theories or constructs. These constructs give you some stable idea of yourself as you navigate new situations, relationships, and
self-revelations. These constructs can be under active revision, giving
you greater or less degrees of freedom to adapt to new challenges. In

the last chapter we visited familiar terrain of psychology, the concept of stable traits. We turn now to a perspective that challenges this position and explores the more mutable aspects of free traits.

MUTABLE SELVES, MYTHIC TRAITS?

When I stepped up to the platform to deliver the keynote address in Arizona, a familiar "click" occurred. I switched from my natural (biologically) introverted personality to something very different. At 8:35 in the morning audiences do not really want to hear modulated, soft-spoken, tentative, introvert-speak, especially after a long bout of impassioned drinking the night before. Even the introverts want something to raise their levels of arousal and get them engaged. So, as a member of the audience, if you were asked a few minutes into my presentation what Professor Little is like, you probably would have said he is a flaming extravert. But I knew better. Or did I? Professor Sam Gosling at the University of Texas raised precisely this question in describing one of his colleagues.

> When there's a disagreement between the self and others, it can be because it's a blind spot and you can't see yourself as you really are. But it can also be the sign of a personal spot—an area where you see yourself more accurately than others do. Take Brian Little, a professor who taught a legendary class on personality psychology at Harvard. According to those who saw him lecture, he was eloquent and garrulous, brimming with ebullience and energy. Unsurprisingly, he was widely known by his students as a raging extravert. Yet Little disagrees. He insists it's all an act executed in the service of being a good teacher. Should we believe him? Isn't it possible, after all, that extraversion is a blind spot of his?[1]

Me? Blind? Possibly. But Sam is a friend and knows me well, and he went on in his article to explain that aspects of my own personality help explain why my students and audiences may easily misconstrue me. I am certainly not rare in this respect. There are times when many of us engage in behavior that leads others to infer, incorrectly, certain "fixed" or stable traits of our personality. Technically, psychologists refer to this as *counter-dispositional* behavior. I have a theory about why and how people act in this way and why it has major consequences for our well-being.[2]

Here is the essential argument. Human personality has both an inner and an outer reality. The inner reality consists of what we are intending to do—what personal projects we are pursuing at any given time. The outer reality consists of images that we create, consciously or not, for others. It is in the nexus between these two realities that our personalities are constructed, challenged, and reconstructed. When we explore this nexus all sorts of strange behaviors can be observed. Neurotics, trying to appear stable, may "leak" their neuroticism, such as when a genuinely nice guy in the bar acts like a complete jerk because he is redressing a painful insult to his partner from the night before. And an introverted Harvard lecturer will appear as a "pseudo-extravert" when he's "on." But, as Gosling goes on to note, after class that same lecturer could be found hiding out in the restroom, lowering his stimulation level. What is going on here, and why does it have important consequences for our well-being?

In the last chapter we explored how stable traits of personality are linked to happiness, health, and accomplishments—the stuff of well-being. In this respect it is logical to conclude that stable individual differences in personality are real and consequential. But some of you may be skeptical about this approach to understanding personality. Doesn't it all depend on the situation? Doesn't context matter? Isn't the notion of stable traits of personality a myth?

The answer in each case is "yes," but with strong qualifications. Obviously we are more likely to be extraverted at a party than at a funeral and more conscientious when doing our taxes than when texting our friends. But trait theory doesn't disagree with this. The *average* level of trait expression does indeed differ from situation to situation. But the *rank order* stability of trait expression—where you stand relative to other people on a particular trait in a given situation—is impressively stable.[3] The class clown in grade school will have downgraded her mischievousness thirty years later, just as most of her classmates have also lowered *their* level of extraversion. But at the class reunion she is still a clown—a classier one, a more measured and mature mischief maker, to be sure, but still an over-the-top extravert. Although everyone, even you, will tilt in the direction of extraversion at a party, the true extravert is more likely to be *exceptionally* outgoing, overly garrulous, and, perhaps, from an introvert's perspective, borderline obnoxious. Introverts, relatively speaking, would be engaged in a more modulated extraversion, relatively outgoing but hardly exuberant. The most introverted of all, however, may simply send a message saying that he will not be able to make the party that night—or any night in the conceivable future.

But you may now have different concerns about trait theory. You might grant that there are some predictable features of a person's traits and that the gist of their various behaviors may be consistent with the notion of relatively stable and enduring traits. But aren't there times, you might ask, when people act in ways that are completely at odds with their basic natures? What do we say about the introvert who isn't just mildly extraverted at a party but truly rocks? Or how about an exceptionally disagreeable person who, for five days over Thanksgiving, is so stable and sweet that her family wonders what's wrong? Let's look more closely at these two examples. Markus is the extraverted introvert; Stephanie, the sweet-and-sour publisher. Each bears a striking resemblance to people I have known for roughly twenty-three years.[4]

Markus is a complex man. He is famous in the indie music business in Montreal as a *bon vivant*, a thrill seeker, and an irrepressible extravert. He is both a musician and an impresario, and he can be banked on to entertain and enliven a room. When he walks in the door the buzz begins. But there is another side to Markus. He can frequently be seen escaping the spotlight, seeking solitude, squirreling himself away reading serious philosophy books and acting like a bona fide introvert. After working the room on a snowy night in Old Town Montreal, he is walking alone in the back alleys, buzzed out, and burnt out. Who, really, is Markus?

Stephanie is a scary woman. Her colleagues in the publishing industry in Manhattan describe her as tough, acerbic, and disagreeable to a fault. She agrees she is disagreeable and actually takes some pride in this depiction of herself. But there are times when she too has been caught in conduct that seems to contradict her reputation—acts of kindness and moments of tenderness that seem totally out of character—like her behavior this Thanksgiving. Who is the real Stephanie?

THREE WAYS OF BEING NATURAL

Biogenic Sources

We can think of our everyday behavior as expressions of three different motivational sources that energize it. The first is *biogenic*: its roots are genetic, and its influence arises from brain structures and processes that the rapidly emerging field of personality neuroscience is studying.[5] Biogenic motives arise from the dispositions and temperaments that a Markus or a Stephanie brings into the delivery room at birth. Such features of personality can be detected in the neonatal ward. If you make a loud noise near the newborns, what will they do? Some will orient toward the noise, and others will turn away. Those who are attracted to the noise end up being extraverts later in development; those who turn away are more likely to end up being introverts.[6]

One of the more interesting ways of informally assessing extraversion at the biogenic level is to do the lemon-drop test. There are several variations on the test, and I draw here on a demonstration procedure I frequently used with my undergraduates.[7] Here are the ingredients you will need: an eyedropper, a cotton swab (the little stick with a wrap of cotton on either end we use for babies and are admonished not to stick in our ears), a thread, concentrated lemon juice (regular lemon juice won't work as effectively), and the willing tongue of a volunteer (such as yourself). Attach the thread to the center of the double-tipped cotton swab so that it hangs exactly horizontal. Swallow four times, then put one end of the swab on the tongue, holding it for twenty seconds. Then place five drops of the concentrated lemon juice on the tongue. Swallow, then place the other end of the swab on the same portion of the tongue and hold it for twenty seconds. Then hold up the swab by the thread. For some people the swab will remain horizontal. For others it will dip on the lemon juice end. Can you guess which? For the extraverts, the swab stays relatively horizontal, but for introverts it dips. The reason is that introverts, because they have relatively high levels of chronic arousal, respond more vigorously to strong stimulation, like lemon juice, so they create more saliva. Extraverts, being less responsive to high levels of stimulation, stay relatively dry mouthed. In fact, there is evidence that because of this tendency toward lower salivation levels, extraverts actually have higher levels of tooth decay than do introverts.[8] I have done this exercise on myself a number of times, and each time my swab dips deeply. I am, at least by this measure, a biogenic introvert. I also suspect that Markus would salivate with great vigor, right on cue.

Each of the Big Five dimensions of personality can be assessed in terms of its biogenic roots.[9] For example, there is growing evidence that highly agreeable individuals have higher levels of oxytocin, a neuropeptide released during birth, breastfeeding, orgasm, and other aspects of intimate behavior. Oxytocin levels can be assessed through blood or saliva assays, and there is currently a great deal of research being done on the gene associated with oxytocin regulation. Consider a recent intriguing study by Alex Kogan and his colleagues.[10] Couples were shown into a Berkeley laboratory to discuss, with their partners, matters that concerned or distressed them. These sessions were videotaped. Participants were also assessed for their possession of a particular variant of the gene that controls the expression of oxytocin. Strangers were then asked to evaluate twenty-second snips of the videotapes and asked to rate how attentive and sympathetic the participants appeared to be when listening to their partners. Participants with the variant of the oxytocin gene were significantly more likely to be rated as showing the sympathetic and agreeable dispositional characteristics.

Theoretically, then, we would anticipate that the biogenically dis-agreeable Stephanie would appear cold and unresponsive when she was listening to her ex-husband talk about his problems (yet again). She also wouldn't be particularly warm and compliant if we asked her to spit into a test tube and have it tested for the absence of that oxytocin gene variant.

The tendency to act in ways directly influenced by such biogenic factors can be fairly said to comprise a "natural" response. It seems reasonable to say that biogenic extraverts and unpleasant people are being natural when they express their extraversion or their unpleas-antness in their behavior. But this is not the only way in which we can act "naturally."

Sociogenic Sources

How we act can also be deeply influenced by *sociogenic* sources that arise through the course of socialization and the learning of cultural codes, norms, and expectations. Behavior with a strong sociogenic source may flow relatively effortlessly because it has been reinforced throughout a person's life as being the appropriate thing to do in various circum-stances. Introversion and extraversion as styles of behavior have this strong sociogenic aspect to them as well as a biogenic aspect. Different cultures place differential emphasis on the importance and acceptability of extraverted behavior.[11] Extraversion, for example, is highly valued in American culture. Part of the appeal of Susan Cain's *Quiet* is that it identifies this inherent cultural bias and calls for an expanded set of options for finding a fit between biogenic and sociogenic factors in the development of personality.

In contrast with the American extraverted ideal, other cultures place a higher premium on introversion. For example, the norms of some Asian countries encourage children not to stick out unduly from the rest of the group but rather to quietly blend in. From an extreme West-ern point of view this is seen as a kind of whack-a-mole perspective that inhibits those who venture out and rewards those who keep their heads down. Such norms have major consequences for intercultural communication.[12]

Imagine what happens when a highly extraverted group of American negotiators are in meetings with a highly introverted group of Asian negotiators. The prospects for successful negotiations may be severely compromised because of the different norms for how to stand, gesture, and express ourselves. High-level negotiators are very aware of these obstacles to communication and take well-designed workshops on how to interact with individuals and groups from other cultures effectively. Naturally the same kind of course is being offered to the negotiators from the "other culture." Americans learn to interact like Asians; Asians learn to interact like Americans. The results can be bizarre—a group of polite, formal, reserved, and reticent Americans interacting with back-slapping, psyched-up Asians doing a pony dance, Gangnam-style, around the negotiating table. Both goodwill and deep confusion are likely to ensue.

Similar cultural differences can be found for the other Big Five traits, such as agreeableness and conscientiousness. There are cultures in which complaining is normative and others in which the norm is to "suck it up and remain civil." There are some cultures famous for their dogged pursuit of goals and others that encourage us to just relax, chill, enjoy ourselves, and face the day with happy shiny faces. In important ways such sociogenic aspects of our conduct are as "natural" as our biogenic tendencies. The influence of culture is profound and pervasive. There are rewards for adhering to cultural scripts and costs for failing to show fidelity to social conventions.

Our first (bio) and second (socio) natures may be in conflict. A biogenic tendency to be assertive and stand out in the crowd may conflict with a cultural norm of "blending in quietly" or our parent's exasperated plea to "grow up and stop embarrassing the whole family." In contrast, if the same biogenic tendency is lodged in someone whose family motto is "Go for it—be awesome!" such audacious acts are less likely to lead to censure than to an enthusiastic round of family high fives.

Let's consider Stephanie and Markus again. Aspects of Stephanie's personality can be explained by the sociogenic influences she grew up with. She is from a culture that places a high premium on assertiveness and holding one's own, even when others might see such behavior as offensive and disagreeable. She learned this early in life and practiced it assiduously when her family arrived in New York. Given her biogenic disposition to be disagreeable, the sociogenic influence simply amplified her tendencies in this direction. And this is why her seeming sweetness with the family over the holidays seemed particularly surprising.

Markus's personality can also be explained in part by sociogenic influences. Markus was adopted at three months of age by a large, loud, effusive, and extraverted family of French Canadians. Yet he was a biogenic introvert, so, unlike Stephanie, he had to find a way to incorporate both sources of influence into his developing personality.

Idiogenic Sources: Personal Projects and Free Traits

Beyond the influence of the biogenic and sociogenic sources of motivation there is another compelling influence on our daily behavior that I call *idiogenic motives*.[13] They represent the plans, aspirations, commitments, and personal projects that we pursue in the course of daily life. Their origin is idiosyncratic and singular. By invoking biogenic causes we can explain a person's behavior as the natural playing out of traits. By invoking sociogenic causes we can explain the same behavior as the natural consequence of social norms. But by invoking idiogenic causes we seek the reasons *why* a person is engaged in a particular pattern of behavior. What personal goal was the guy in the restaurant trying to achieve by repeatedly sending back the steak? What was the aspiration that animated the behavior of Markus in Montreal that particular winter? What commitment might have explained why Stephanie resists both her biogenic nature and her sociogenic nurturing on this particular holiday evening? To answer these questions, we need to understand some more about personal projects, and we need to introduce the notion of "free traits."

Personal projects are the stuff of everyday life. They can range from the very trivial pursuits of a Thursday morning (e.g., "put out the dog") to the overriding aspirations of our lives (e.g., "liberate my people"). In Chapters 9 and 10 I provide details about how personal projects play a direct role in enhancing or frustrating our well-being. For now, I want to focus on how being engaged in projects that matter to us can lead us to act in ways that surprise others and, sometimes, ourselves.

Take Stephanie. We know that she is typically a highly unpleasant person. She scores low on the relatively stable trait of agreeableness, and she is culturally attuned to being rather combative. Both her colleagues at the publishing house and her friends and relatives at home are used to her abrupt demeanor, as was her ex-husband. So at Thanksgiving this year Stephanie's surprisingly sweet behavior surprised everyone.

What her family did not know at that time was that Stephanie was heading off to Australia right after Thanksgiving, where her publisher had asked her to head up the management of a new venture. This was to be a three-year commitment with minimal possibilities for visits back and forth to New York. Her daughter is now six months pregnant, and Stephanie has been thinking about having her first grandchild arrive while she is away knocking heads together in Sydney. She has also been reflecting on her son-in-law and his parents, all of whom will be at the Thanksgiving gathering. They are quiet, thoughtful, and very sweet folks, and Stephanie knows they find her a chore to keep jolly on festive occasions. After reflecting further, Stephanie decides to make some changes in her relations to her family. Though it is still only vaguely formulated, she creates a personal project of "being a more nurturing mom." This in turn impels her to act in a supportive and pleasant manner. Someone who had no previous experience of Stephanie who observed her at the Thanksgiving table that year would evaluate her as very high on the Big Five trait of agreeableness. I call such behavior the enactment of a *free trait*, in contrast with relatively fixed traits.

Markus too was pursuing a personal project that led him to exhibit free trait behavior. Despite his biogenic introversion, Markus had a driving passion to be a music producer. This didn't require much of a stretch for him; he was able to be absorbed in his music and remain rather oblivious to others around him. But when he discovered a talent for promoting gigs and producing indie records, he increasingly found that his social commitments, sometimes extending into the very wee hours of the morning, were becoming increasingly demanding. Few suspected that he wasn't the irrepressible extravert he appeared to be, but it is better to describe him as a *pseudo-extravert*, someone who is adopting a sociogenic script to promote a personal project that mattered deeply to him, despite his biogenic introversion.

Why do people engage in free trait behavior? There are many reasons, but two are particularly important. We enact free traits out of professionalism and out of love. Stephanie, for all her disagreeableness, loves her family deeply, and by acting out of character is able to express that love more effectively than if she surrendered to the default option of being her biogenic self. Markus is a consummate professional, and one of the requirements of his role is to engage with and inspire fellow musicians and their patrons. It is precisely this professionalism that he regularly displays despite his biogenic disposition to disappear quietly into the background. Such behavior defines him and sustains his reputation in the music community. It is his defining feature: the singular mark of Markus.

ACTING OUT OF CHARACTER

Given these distinctions between the three motivational sources of daily action, what does it mean to say that a person is acting naturally? Actions based on our biogenic dispositions are clearly natural in the sense that they directly reflect our biological needs and stable preferences. Stephanie interrupts a colleague with a withering comment, and this isn't seen as surprising. It's *natural* for Stephanie to act this way. But does this mean that we should regard her Thanksgiving behavior as unnatural, insincere, or disingenuous? Not necessarily. Her enactment of a core personal project that entails a shift away from her traditional ways of acting can be said to be *acting out of character*.[14]

This phrase can be interpreted in two different ways, both of which I want to retain. One means acting *away from* what we normally expect. This is one of the meanings implied when we say that Stephanie and Markus are acting out of character. But the phrase "acting out of" can also mean "acting because of," as in, "He sent the steak back out of spite" or "She did it out of compassion." So when I use the phrase "acting out of character" it means two different but equally powerful ways of explaining a pattern of behavior. It simultaneously means people are acting inconsistently with what we have come to expect *and* that they are doing it because of something in their character, because of the *values* they wish to express.

Here's an example. Imagine you are a mother of a six-year-old daughter and you are holding a birthday party for her with fifteen of her friends in attendance. Imagine too that you are a rather anxious introvert (biogenically speaking) but that you have a deep desire to put on a great party for your daughter. This is a core value for you. It is, admittedly, difficult for an introverted mom to play Pin the Tail on the Mommy without incurring some strain. But you do it, and everyone has a blast. Is this disingenuous? No. Is it faking? Not at all. But some of the other parents who come to pick up their children at four o'clock may comment on their way out that you were definitely acting out of character (in the first sense). At PTA and community functions you seemed always to be the quiet, subdued person, but this afternoon you have transformed into a whirling dervish, delighting the kids and surprising their parents. But you were also enacting a personal project—"give my daughter an awesome birthday party"—that was based on a core value of being a good mother. So you were also acting out of character in the second sense.

This takes us back to the question of being natural. What would we conclude when we see a man bellowing out his inimitable version of "Do You Think I'm Sexy" at a karaoke bar? We might assume that this is the way he is and that he is behaving naturally; that is, we assume he is a biogenic extravert. He is, in our terminology, showing fidelity to his biogenic traits. But isn't it equally plausible to say of a person who is sociable and demonstrative at a kid's birthday party, even though we know she is biogenically introverted, that this too is only natural? It's her much-loved daughter's birthday, for goodness sake, and there are very well-known sociogenic scripts she can adopt to express that love. It's only natural to do so. So the position I take stresses three potentially conflicting forms of fidelity—fidelity to one's biological propensities, to one's cultural prescriptions, and to one's core personal projects. Each of these is *natural* in its own compelling way, and the way they are artistically choreographed in your life has important implications for your health and well-being. So it is important, as you reflect on your life, to ask three questions: What do you gain by pursuing personal projects and enacting free traits? What are the dynamics of acting out of character? And what might be the costs?

The Benefits and Costs of Acting Out of Character

The great benefit of adopting free traits is that they can advance the personal projects that bring a sense of meaning to your life. It is possible—though unlikely—that simply allowing our biogenic natures to direct our lives will be a satisfactory strategy. It is true that those who have biogenic traits of openness, conscientiousness, extraversion, agreeableness, and stability may fare very well in societies in which these dispositions are valued, and those at the opposite end of the trait spectrum are likely to have less than happy lives. But life throws out challenges that require each of us at times to shift our orientations, to defy our biogenically fixed traits and adopt free traits. Another benefit that arises out of engaging in free trait behavior is that it expands us. Stephanie is stretched when she is able to act sweet and fulfill a commitment; Markus is more Markusian when he works a room, clinches a deal, and transforms the indie music scene.

There is, however, a possibility that engaging in free traits or acting out of character can actually take a toll on us. To understand why, we need first to know something about how stable traits and free traits play out over time dynamically.

Free Trait Dynamics: Getting Fit, Sucking It Up, Letting It Out

I believe that *protractedly* acting out of character through free traits can extract both psychological and physical costs. Evidence from several research literatures provides some relevant evidence in support of this proposition.

The situations and settings of our everyday lives play an important role in the quality of our lives. The better the "fit" between a person's biogenic traits and the characteristics of the environment, the better the consequences for well-being. One of the functions of the environment is to provide the right resources for enabling our personal projects. Our own research has shown that highly sociable individuals are happier if they are engaged in personal projects that involve a lot of social interaction.[15]

A striking example of looking for a match is the experience of Peter, an old student of mine. In my psychology laboratory class at Oxford in the late 1960s he took a couple of the personality scales we have discussed in this book. He told me that on the extraversion scale he had as extreme a score as one could obtain. But prior to coming to Oxford Peter had lived as a monk for some time in an isolated Belgian monastery, where he had taken a vow of silence. In terms of having a good match between biogenic disposition and environment, this was not a very promising vocation! From a free trait perspective I would have predicted that his need to act counter-dispositionally over a long period would have been exhausting for him, and he would have simply petered out. I am pleased to say that he eventually ended up as a highly successful professor of education, in which his need for connection and interpersonal excitement fit very nicely.

So getting a good match between our biogenic dispositions and the contexts of our daily lives should enhance our performance and well-being, and having a major "mismatch" might put us at risk. Highly disagreeable people, for example, are more likely to thrive as bill collectors than as counselors, and those who are open to experience will find a better match for their dispositions in some of the "villages" in New York City than in the suburbs in southern North Dakota. (However, do see Chapter 8, where we talk about a surprising Fargo factor.)

But what if your environment simply doesn't supply the resources to satisfy your disposition and facilitate your projects? One can, like Peter, leave that environment and start a whole new way of life. You could also modify an ill-fitting environment by creating microniches within it, although trying to institute Flash-mob Fridays in a Benedictine monastery would be a challenge. But you can also do something that gets to the very heart of the notion of free traits—you could change your own personal style. A study that casts a most interesting light on this possibility posed the following question: Do students change on the Big Five personality dimensions as they progress through the four years of undergraduate education? In a longitudinal study of the students at the University of California, Berkeley, researchers found that there were indeed significant personality changes. Perhaps surprisingly, the students became increasingly disagreeable as they progressed through their programs. They also became less neurotic. What kind of change is this?[16]

I think this shift reflects an accommodation to a highly competitive and demanding academic environment in which a premium is placed on the ability to critique and challenge convention dispassionately. It is possible that the survey results reflected deep, *substantive* personality change among some students. But from a free trait perspective the changes the survey measured were more likely to have been *strategic* than substantive biogenic ones—more the enacting of a free trait than a shift in an enduring trait. The personal projects driving such change would be pursuits like "impress my seminar professor" and "get a great recommendation for grad school."

There is another implication if we interpret these results from a free trait perspective. The greater the discrepancy between the biogenic traits and the free traits, the more difficult this transition is likely to be. Highly emotional and agreeable first-year students at Berkeley will find the upper-division transformation more challenging than will those who had already entered pretoughened to be calm, cool, and critical.

The ability to suspend one's biogenic tendencies in order to adapt to situational demands is also reflected in another study of university students who were assessed during their first semester at college.[17] Students generated a list of their personal projects that researchers then categorized into life tasks typical of that stage of development. Two major life tasks for first-semester college students are doing well academically and creating a new and rewarding social life, or, as Bob Hogan calls it, getting along and getting ahead. Which of these two personal project priorities best predicted success and well-being during the first semester? The result showed that it is not a matter of one priority versus another; rather, there is a *dynamic* process going on—it all depends on the timing. Students who started out investing in social projects but were unable to shift to academic priorities did not fare well that semester, nor did students who started off and persisted with an exclusive focus on academics. Students who flourished during their first semester were those who were able to give an initial priority to social tasks but then switch to academic ones. Presumably during the stressful moments of intense academic pursuit later in the term those who had created their social networks early on could draw on them as valued resources. Notice again that the tendency to give priority to academic or social projects is likely related to biogenic traits—conscientiousness and openness in the case of academic tasks, and agreeableness and extraversion in the case of social tasks. This suggests that gregarious, friendly, outgoing students will cope well early in the term when establishing a network of friends, but they will need to suppress those extraverted tendencies when the academic crunch begins a short time later. If we walk on campus late at night in mid-October, we would likely look up to see them in the library acting out of character as pseudo-introverts, missing their friends, hitting the books, and making their mark.

Sucking It Up: The Price of Pretending

Have you ever wondered whether flight attendants have been getting grumpier over the past decade or so? For years the airlines' training manuals required their flight attendants—then called stewardesses—to manage their emotions and literally put on a face: mandatory smiling was a professional requirement. No matter how hassled, tense, or grumpy they might feel on a particular day, they needed to suck it up and put on the Pan-Am smile, together with Revlon Persian Melon lipstick, sky-high heels, and the solemn promise not to shrink to less than five feet two inches in height. In recent years these restrictions have been lifted, although there remains, at least with some airlines, an expectation that flight attendants will dole out a few smiles along with the delicious pretzels and instructions on how to insert the flat metal end into the buckle. For naturally affable, outgoing flight attendants, the professional requirements—the sociogenic demands—provide a fit with their biogenic natures, and there should be very few negative consequences. But for those who need to suppress their biogenic traits, there may indeed be costs.[18]

There is an intriguing research literature that raises some important warnings for individuals who protractedly engage in free traits through suppressing their biogenic traits. The central idea is that suppression causes arousal in the autonomic nervous system, and if such arousal becomes chronic, it can extract a health cost. Jamie Pennebaker and his colleagues have shown, for example, that students who have suppressed something important about themselves, such as deeply unpleasant events from their childhood, have chronically raised levels of autonomic arousal and have more health problems than do those who are not suppressing something important.[19] They have also shown that if you open up about the suppressed aspects of your life by writing or talking about them, something interesting happens to autonomic arousal. First, when opening up, the arousal level briefly increases—it isn't easy to talk about that which you have been suppressing. But after opening up, arousal diminishes and not only goes back to the prior level of arousal but actually is lower than it was before the opening up. Those who have opened up are healthier, and this is in part due to enhanced immune system functioning.[20]

I am suggesting that this may be what happens when we engage in free trait behavior over a long period of time. A biogenically agreeable woman who is required by her law firm to suppress her pleasantness and act aggressively may experience signs of autonomic arousal—such as increased heart rate, sweating, muscle tension, and a stronger startle response. If the culture of the law firm is that you simply do not talk about such matters, that it would be unprofessional to vent, the costs will be particularly taxing. And Markus, the musical impresario, may be exhausted in the back alleys of Montreal precisely because he has never been able to confide to anyone that he desperately needs a break from the nonstop buzzing conviviality of the music scene.

There is another twist that derives from suppressing our biogenic natures. The fascinating research of Dan Wegner provides compelling evidence that suppressing a thought—his classic example is not to think about a white bear—stimulates what he calls "ironic processes."[21] Suppression—explicitly not thinking about a white bear—requires that we have a representation of that which we are trying to suppress, which leads to a state of hypervigilance, depletion of cognitive resources, and the ironic reappearance of that which was to be suppressed. In short, not thinking about the white bear entails thinking about the white bear that we are not thinking about. (If you are getting frustrated that you can't get a white bear out of your mind as a result of reading this, let me suggest that you not think about a green cat instead.)

I am suggesting that the same process takes place when we are explicitly attempting to act in a counter-dispositional manner, when we are, in other words, engaged in free trait behavior. If Markus is engaged in a particularly stimulating negotiation about an upcoming gig, he might display a slight leakage of introversion—a micromomentary pause in his otherwise forceful negotiating voice, a slight evasion of eye contact with the pit bull entertainment lawyer across from him. Stephanie too may find that despite her commitment to ensuring her family's well-being, she still finds herself flaming her daughter about a suggested name for the baby in a late-night e-mail she now wishes had remained unsent (although little Noah may later appreciate not being called Grimly).

Restorative Niches: Reducing the Costs of Acting Out of Character

Is there anything we can do to reduce the potential costs of acting out of character? One thing we can do is to find ourselves a *restorative niche*, a place where we can obtain some respite from the physiological costs of free trait behavior and can indulge our biogenic "first natures." Here's a personal example.

As a lifelong biogenic introvert, I tend to overload easily and am particularly sensitive to various forms of social stimulation. It's not that I don't enjoy that stimulation; it is that I cannot perform effectively in such a situation. For years I used to visit the Royal Military College in St. Jean-sur-Richelieu, Quebec, where I would lecture military leaders on the fine art and hard science of understanding personality. I would typically drive down the night before and then spend a full day with them, lecturing for three hours in the morning and then another three after lunch. One of my core personal projects is engaging fully with my students, whether they are college sophomores, four-star generals, or the folks back at my talk in Arizona whom we have left dangling for a bit. To fully connect with the audience, my lectures need to be fast paced, intense, and interactive—in short, highly extraverted. So by the end of the morning session at the Royal Military College I would be in a state of hyperarousal, well over the optimal level required for lucid lecturing.

Then it would be lunchtime. Just when I most needed to lower my level of arousal, the officers would invite me to the Officers' mess. Although I did go along with this for a few visits, I soon realized that it compromised the quality of my lectures in the afternoon. So I hit upon a strategy. I asked whether, instead of lunching with the officers, I could take a walk by the Richelieu River that ran alongside the lecture theater. My pretext was a keen interest in the variety of craft that sailed along the Richelieu, but, of course, my main motivation was more strategic. I needed to lower my arousal level. This strategy worked well for a couple of years, but then the campus moved to a different location and the river stayed where it was. So in the subsequent visits I needed to find a new niche, a different place for lowering my arousal level. I found an ideal place: the men's room. I would choose the cubicle furthest from the action and quietly muse on life and my afternoon lecture while restoring my biogenic nature.

One day, unfortunately, my safe place, my restorative niche, failed me completely. I was well into the pleasant state of arousal reduction when I heard what could only have been the sounds of a supernatural extravert. He was the loudest hummer I had ever heard. He burst into the men's room and lurched his way to the door of cubicle two (I was peeking through the slits in my door). He must have spotted and recognized my nonmilitary footwear, because he stopped, turned, and came straight toward my cubicle. I could feel my autonomic nervous system kicking in. He sat down in the cubicle next to me. I then heard various evacuatory noises—very loud, utterly unmuffled. We introverts really don't do this; in fact, many of us flush during as well as after. Finally I heard a gruff, gravelly voice call out, "Hey, is that Dr. Little?" He was an extravert—he wanted to chat! Now if anything is guaranteed to constipate an introvert for six months, it's talking while on the toilet, and of course my arousal level was off the chart. Needless to say, after our extended, animated interstall conversation, I was somewhat less able to mount a lucid afternoon lecture that day. I decided I had to modify my restorative niche behavior slightly from that point on and to find a strategy to avoid being detected. So from then on, if you were keen to chat with me at the break in the lecture, you may not see me in the men's room. But I'll be there, in the furthest-flung stall, lowering my level of arousal—feet up!

TOWARD A FREE TRAIT AGREEMENT

Restorative niches are not just for introverts acting as pseudo-extraverts. Extraverts who are "pseudo-introverts" at work don't need a quiet

hideaway to restore them; indeed, my niche could well be your night-
mare. And vice versa. They need something that will re-engage them—a
throbbing night club experience, possibly with Markus, would do very
well.

Let's go back to the ballroom in Arizona where we began this chapter. After clicking into delivery mode, I began the lecture by giving the
audience a sense of what fixed traits of personality tell us about our
prospects for enhanced well-being, and I did it in full flight as a pseudo-
extravert. But then I told them about free traits, acting out of character,
and restorative niches, and I warned them not to follow me to the restroom when I concluded my presentation. After the lecture I chatted
for a few minutes with some people who had wanted to talk about
applying the information to their families. Then, as I gathered up my
belongings, I spotted a man at the door who was obviously waiting for
the others to leave. He approached me. He was very direct. "Brian, according to your personality test I am exceptionally disagreeable." "Oh,
I'm sure you're not," I said. "Shut up," he interrupted. He told me
again that he was deeply unpleasant and made it clear that no amount
of repudiation from me was going to pass muster. He then proceeded to
tell me that he had just spent two weeks visiting his dying mother. He
was with her every day, and he was soft, loving, and adoring. "Totally
out of character," he said, picking up the language of the lecture. "But
if you are right," he said, "then I could be paying a price. Both my sister
and I are grieving, but she is in much better shape than I am. She is
naturally pleasant, in fact quite sickeningly so. We are both saddened
by Mom's death, but I feel completely burnt out. So my question to
you is, what kind of restorative niche would work for me?" I was intrigued. Although our preliminary research on free traits had focused
almost exclusively on extraversion and pseudo-extraversion, he was
asking about the cost of acting as a pseudo-agreeable person, not just
for a fleeting encounter but for an intense and extended period. I asked
him whether he played recreational hockey—I suspected he might because his jacket had a recreational hockey league logo on it. "Yeah,"
he replied. "Is hitting allowed?" I asked. "Yep," he said. "Well, then I

suspect a few games of rough-and-tumble hockey, where you truly kick some butt, would be mightily restorative for you." "Can I tell the ref it's therapy?" "Sure."

I think there is a moral dimension to free trait behavior. Acting out of character is value driven. We rise to occasions when we might have defaulted to our biogenic selves. We do it out of love and we do it out of professionalism, and through it we deliver on our personal and professional commitments. But it can take a toll on us. To mitigate this toll I propose we develop a "free trait agreement." It isn't a formal document but rather an informal pledge to ensure that each of us has a chance to create and enter the restorative niches that satisfy our biogenic natures. This can be as simple as showing forbearance and support to those whose behavior may seem rather puzzling. If your gregarious wife, after two weeks of concentrated and isolated work, heads off for a wild and wacky weekend with her girlfriends, consider the possibility that she isn't doing it because she doesn't love you; she's doing it for fun and release but also in part because she knows it enables her to love you better when she returns restored to her extraverted self. And that kind and sensitive fellow in Internal Audit who has to be Mr. Poopyhead all week long should be allowed to have his guy hugs on Friday when the fight is temporarily suspended and he has a fleeting moment when he can be truly himself.

I want to end the chapter with a very personal message: don't take your Big Five trait scores too seriously. Don't let them cage you in or leave you, in T. S. Eliot's imagery, pinned and wriggling on the wall like a perfect biological specimen. Don't tell other people your Big Five scores (although, unfortunately, extreme extraverts will have already shouted out theirs for all to hear). You are more nuanced than a single number or five single numbers. *Do* talk about the things you are doing that matter to you in your life—your core projects, continuing commitments, and future aspirations. Once these become the clear focus, your relatively fixed and more strategic free traits are seen in a different light. From a trait perspective you may be a neurotic introvert—fair enough—but such a depiction seems unduly limiting. I believe you have more degrees of freedom than that. By acting out of character and engaging in free traits you can advance the core projects you hold dear.

For some individuals the ability to "click" into different modes of self-presentation and adopt free traits is relatively easy. But for others it makes no sense to be anything other than you. These differences have major consequences for how you express your personality and how you relate to others. The next chapter will help you determine where you stand on this important aspect of personality and why it matters.

chapter four

Mutable Selves:
Personality and Situations

WHY DO SOME INDIVIDUALS SEEM TO BE THE SAME PERSON no matter what the situation is, whereas others shift their self-presentation, chameleon-like, and appear to be different people depending on the situation? How about you? At a funeral do you act funereally? At a barbeque do you really do barbeque or are you actually more funereal, at least in the eyes of those who are throwing buns at each other?

Do these tendencies have an impact on our achievements and well-being? We explore these questions in this chapter. But first it will be helpful for you to take the SM scale, below, or at least read over the items. Despite its name, I can assure you that the SM scale doesn't deal with anything too kinky.

SM Scale[1]

The statements below concern your personal reactions to a number of situations. No two statements are exactly alike, so consider each statement carefully before answering. If a statement is true or mostly true as applied to you, mark T as your answer. If a statement is false or not usually true as applied to you, mark F as your answer. It is important that you answer as frankly and as honestly as you can. Record your responses in the spaces provided on the left.

_____ 1. I find it hard to imitate the behavior of other people.

_____ 2. At parties and social gatherings, I do not attempt to do or say things that others will like.

_____ 3. I can only argue for ideas which I already believe.

_____ 4. I can make impromptu speeches even on topics about which I have almost no information.

_____ 5. I guess I put on a show to impress or entertain people.

_____ 6. I would probably make a good actor.

_____ 7. In groups of people, I am rarely the center of attention.

_____ 8. In different situations and with different people, I often act like very different persons.

_____ 9. I am not particularly good at making other people like me.

_____ 10. I'm not always the person I appear to be.

_____ 11. I would not change my opinions (or the way I do things) in order to please someone else or win their favor.

_____ 12. I have considered being an entertainer.

_____ 13. I have never been good at games like charades or improvisational acting.

_____ 14. I have trouble changing my behavior to suit different people and different situations.

_____ 15. At a party I let others keep the jokes and stories going.

_____ 16. I feel a bit awkward in company and do not show up quite as well as I should.

_____ 17. I can look anyone in the eye and tell a lie with a straight face (if for a right end).

_____ 18. I may deceive people by being friendly when I really dislike them.

SCORING THE SCALE: The scoring key is reproduced below. You should circle your response of true or false each time it corresponds to the keyed response below. Add up the number of responses you circle. This total is your score on the SM scale. Record your score below.

1. F 2. F 3. F 4. T 5. T 6. T 7. F 8. T 9. F 10. T 11. F 12. T 13. F 14. F 15. F 16. F 17. T 18. T

MY SCORE: _____

The SM scale measures *self-monitoring*.[3] High self-monitors (HSMs) are concerned about how others see them, and they behave so as to reflect the norms and expectations of the situations they enter. Low self-monitors (LSMs) are less concerned with how others view them and are guided in their behavior by their own traits and values rather than situational expectations. Knowledge of SM scores provides us with rich material for reflection on personality and well-being. Are our relationships more likely to flourish if we communicate with complete candor, or are subtlety and nuance better options? Can success in our working lives be better achieved by carefully monitoring the social situations we encounter? Or is it better to simply be oneself? Knowing where you stand on the SM scale will help you answer such questions.

Now it is entirely possible that you demurred from taking the SM scale because it was too long or because you are the type of person who, in principle, never takes such scales. There is another way—a much quicker one—of getting a hint as to where you would stand on the scale. As you read this, pretend that I am standing opposite you when I ask you to do the following. Please write, with your finger, the letter Q on your forehead. Do it now. Did you put the tail on the right side or the left as viewed from inside your head looking outward? Which you did might offer a clue about whether you are a high or low self-monitor. Oh, and if you wrote a letter other than Q on your forehead out of sheer defiance, well, bless your disagreeable little heart.[4]

PERSONALITY, SITUATIONS, AND A PINCH OF SALT

Before I review some of what we know about self-monitoring I want to provide some context. In 1968 Walter Mischel, then a professor at Stanford University, published his *Personality and Assessment*, a book that had a dramatic effect on the study of personality.[5] He argued that personality, as traditionally conceived, was a myth. Mischel's analysis of the available empirical research concluded that the conventional assumption—that people are driven by fixed traits that generalize across diverse situations—was untenable or at least required serious reconsideration. Mischel advanced an alternative explanation for our daily behavior, a social-cognitive one that explains behavior in terms of the situations that we encounter and our cognitive processing of those situations.[6]

The ensuing trait debate in personality and social psychology pitted personality psychologists defending trait positions against social psychologists adopting a situationist approach. It was a rancorous debate but one that led to major developments on both sides of the divide. The most agreed upon resolution is that it is the *interaction* between personality traits and situations that best accounts for behavior.[7] Wild parties and quiet conversations will attract extraverts and introverts respectively. Another resolution is to focus on the actions that individuals are engaged in—the projects and tasks that fill their days and occasionally their nights. The traits we have and the situations we encounter both play a role in our daily pursuits, so studying those pursuits enables psychologists to integrate the strengths of both personality and social psychological analysis. That happens to be my own perspective, and I discuss it in detail in the final two chapters.

Professor Mark Snyder, the creator of the SM scale, had another, very creative resolution to the great trait debate. He argued that in their daily behavior LSMs are guided by their traits, whereas HSMs are guided by situations. The distinction proved valuable, and it allows us to bring into focus a diverse set of tendencies and preferences, some of which seem rather surprising, such as how we put salt on our food.

You are about to take a bite out of a steak (or, if you insist, a large slab of tofu) that has just appeared on your plate. Do you taste it before putting on salt? As part of his doctoral research at Stanford, Snyder examined this very question and found that individuals who scored high on the self-monitoring scale were more likely to sample the steak before salting.[8] Those scoring low on the scale were more likely to salt it before tasting it or not to put on salt at all. It is as though low self-monitors know their salt personalities very well and act accordingly, whereas high self-monitors need to check the situation, in this case the flavor of the steak itself, before chomping down. The behavior of the LSMs, according to Snyder, is consistent with their deeply rooted general tendency to rely on themselves rather than the situation to guide behavior.

Having started with steaks and condiments, now we can expand the food metaphors to help us understand the impact of self-monitoring dispositions on how we think about ourselves. Are we better described as onions or avocadoes? When asked to list their attributes, HSMs tend to report more publicly visible or available aspects of themselves such as physical features, status, and roles they play. LSMs are more likely to report their internal attributes such as their values, enduring preferences, or the kind of traits subsumed by the Big Five factors. When researchers examine these self-concepts it appears that the HSMs are rather like onions—one keeps peeling back layer after layer until one discovers no substantial self there at all.[9]

Perhaps your coworker Elizabeth is like that. You are never really sure where she's at, nor, you worry, is she. Her self is mutable and differentiates into many subselves. There is no Elizabethan essence to Elizabeth. Conversely, LSMs are more like avocados: when you dig down you discover a pit, a firm core that is invariant. Maybe your friend Doug is an LSM—Doug is always Doug, he never plays with being Dougie the Whimsical; he doesn't do Douglas the Serious. He's just plain Doug. And with him you know what you are going to get. His core is solid and his self is immutable. Some might call it rigid.

In the previous chapter we talked about how individuals can adopt free traits to help advance core projects that matter to them even when it means acting in ways that are discrepant with their biogenic selves. HSMs should be particularly adept at such acting out of character, whereas LSMs are more likely to be puzzled as to why they should act that way at all.

In the following sections I explore some of the ways in which self-monitoring orientation influences our lives, from our friendships to our career trajectories. At the end of this exploration we will be faced with a question about values and character—questions about the way we *ought* to live our lives. Science—in this case personality research—is not designed to adjudicate such questions, but it can help us reflect more deeply about questions of value. To sharpen your personal sense of these differences it might help to answer two questions. First, would you prefer your romantic partner to be an LSM or an HSM? Second, which would you prefer the leader of your nation to be?

ORCHESTRATING THE SITUATION

Given their disposition to be attentive to situational cues, HSMs are keen to make sure they know the nature of the situations they are about to encounter. The *clarity* of the situational expectations is particularly important to HSMs. This was nicely demonstrated in a study in which students were given the choice of entering or not entering a situation in which they had to behave as extraverts. HSMs were far more likely to enter if the situation was defined clearly, irrespective of their own extraversion level. However, LSMs' choices were based on whether they were introverts or extraverts; if they were LSM extraverts, in they went. Also, when asked how the situation might be changed to make them more willing to enter it, HSMs transformed it so as to provide clearer guidelines for conduct. LSMs transformed the situation to more closely match their own dispositions to be introverted or extraverted.[10]

First, Google

Given this line of research, I suspect that HSMs have a particular fondness for Google. It allows them to do due diligence on the important situations they are likely to find themselves in before they enter them and to gain clarity instead of a series of question marks. Consider a job interview. Most applicants will seek out information on the nature of the company they are hoping to join. But with an HSM, it goes much further. I have known HSMs who will Google not only the details of the company but also the bios of the people interviewing them to find out where they went to school and even what their hobbies and social networks look like. Then, when they are being interviewed, they can guide the conversation in directions in which they can connect more closely with their interviewers: "Ah, yes, Mr. Thompson, that sounds like a question that a sociology graduate from Brandeis would ask." The problem with this, of course, is that it can be construed as creepy behavior, particularly so if the interviewer is an LSM and actually went to McGill.

LSMs do not need to worry unduly about how to dress, speak, or express themselves because their default option—their preferred option—is to draw on their own traits, preferences, and beliefs. HSMs, as we have seen, desire clarity in the situations they confront. Certainly, in my experience, they find it stressful to respond to requests to attend functions that do not provide a very clear script as to what is going to happen and how to behave. Imagine a colleague phones you and asks whether you are free for a dinner party tomorrow night. An LSM is likely to attend or not attend depending on her evaluation of the person inviting her. An HSM wants to know such things as who else is coming, is it formal or informal, how long will it last, should I bring something, and what is the real reason for the dinner? Unfortunately for the HSMs, these are not easily Google-able questions.

Let's take another example. At this very minute, right now, my research assistants are outside your door wanting to come in and look at your living space. How would you feel and what would you do in the short time you have until they come in? Now, consider again your self-monitoring score. There is a tendency for LSMs to be nonplussed by such an event. Their place is a reflection of themselves and their traits and preferences, and they have no desire to have it any other way. HSMs, however, would likely be perplexed. They would want to arrange the room to accord more with the desired image, or at least not to give the appearance of being total slobs. For an HSM, a night from hell would comprise a Thursday evening at home during which there is a knock on the door and, with no prior notice, in tumble her current boyfriend, past boyfriend, third-grade teacher, divorced parents, Professor Little, and Wolf Blitzer. For LSMs? No problem. Come on in, everybody! Seriously.

You're Inviting Her? Self-Monitoring and Activity Partners

In their relationships with others there is a predictable tendency for HSM individuals to be highly sensitive to the fit between the situations or contexts they confront and the friends or partners they choose to be with in that situation. Imagine, for example, that you have to choose a friend with whom to go to each of two different functions. One is a football tailgate party in Tuscaloosa, Alabama, and the other an after-the-ballet soiree at the Juilliard School in NYC. You have two friends who come to mind: one a University of Alabama fan who knows more about good beer and Crimson Tide football than anyone really should, and another who is a cellist in New York City and is dating a Juilliard student. HSMs most certainly—and probably even LSMs if pushed—would be able to say which friend would fit best with which event.

But let's make it harder. Let's say that despite the fact that both are your friends, you like the Tuscaloosa football friend more. Would this make a difference to your choice of whom you would take to each event?

The experimental research suggests that it would: LSMs would choose the football fan for *both* events, whereas HSMs would select the "appropriate person" for each event.[11] The thought of being accompanied by a Crimson-faced football fanatic at a sophisticated soiree would be rather disconcerting to HSMs, as would be hanging out with a cerebral Cedric at a beer tent in the Quad. For LSMs, you go with whomever you like most. Clearly these differences in self-monitoring disposition can create interpersonal friction, especially when friends become lovers, as we will now see.

Romantic Relationships: Commitment and Flexibility

If the choice of activity partners can give rise to interpersonal challenges, the difficulties are even greater when we consider romantic and intimate relationships. When given biographical information and photographs of potential romantic partners, HSMs pay more attention to the photographs, whereas LSMs spend more time focusing on the biographic information. This is consistent with other evidence that physical appearance and cues of social status are more important factors in HSM preferences in partners, whereas personality and values matter more to the LSMs.[12]

Self-monitoring dispositions also have consequences for romantic relationships' stability. LSMs tend to have more enduring relationships and are less likely to divorce or to engage in extramarital affairs than are HSMs. To put a positive spin on it, HSMs could be seen as highly flexible in managing romantic interests, although in the eyes of their LSM partner, the term "flexible" might not be the first that comes to mind. At the very worst, HSMs could be like Cecilia in Paul Simon's song who, when her lover gets up to wash his face, lets someone else in bed to take his place. This is not very reassuring behavior.

This is certainly not to suggest that HSMs are all disloyal philanderers; rather, HSMs readily adopt a style that best fits the situation at hand, and sometimes this means a lack of constancy, particularly if viewed from the perspective of an LSM. William James famously proposed that we have as many social selves as we have others about whose opinions we care. But we can now add the qualification that this holds *particularly* for HSMs. James's generalization is less likely to apply to LSMs, for whom there is a unitary self, a solid core, which resists being distributed into subsidiary, specialized selves.

In discussing these results with my students over the years I became aware that the behavior an HSM may call "socially appropriate" an LSM may call "being a fake." This one of the major sources of frustration for couples who differ in self-monitoring orientation. The typical example they raised was what happened when they would visit each other's families over a holiday period. At such gatherings the HSM is likely to adopt a different style of interacting with different people, whereas the LSM is more likely to act in the same way with everyone. LSMs would express frustration because of the ease with which their partners seemed to switch attitudes, preferences, and beliefs in conversation with different people. If the family gathering includes a real diversity of people, the HSM can, during the same evening, appear to be very conservative to your Tea Party father, a left-tilting liberal to your very alternative uncle, and awesomely cool to your younger brother. So with which of these alternative selves were you falling in love? Isn't it risky to commit to one whose self is so mutable? But the HSM also expresses grounds for frustration. Can't your romantic partner simply go with the flow of the situation and flex a bit instead of alienating two-thirds of your family? Might he actually be, as others have hinted, a self-focused and insensitive jerk?

I'm Out of Here: Commitment, Success, and Organizational Life

There is empirical evidence that HSMs have greater occupational success than do LSMs. In work groups HSMs are more likely to emerge as leaders, and they also receive higher performance ratings in management positions that involve "boundary spanning," that is, those requiring attention to diverse roles and social cues.[13]

Some of the HSMs' skills are very nuanced. When they are responsible for the failure of a work project, HSMs are more likely than LSMs to rationalize their actions and control the flow of information conveyed to others about the failed project. LSMs, in this respect, can find themselves receiving more censure for projects that go astray because they fail to spin the narrative to deflect attention away from themselves. Many people, particularly other LSMs, will likely regard such behavior as refreshingly honest. But that straightforward directness and lack of self-posturing does not always lead to smooth interpersonal functioning in organizations. When dealing with workplace conflict LSMs are more likely to be forceful and one-sided (from their perspective, the right side). In contrast HSMs are more likely to resolve conflicts through compromise and collaboration.

In an article entitled "Do Chameleons Get Ahead? The Effects of Self-Monitoring on Managerial Careers," Martin Kilduff and David Day report a longitudinal study of a cohort of MBA students who recorded how their careers progressed over a five-year period after graduation.[14] Early in their MBA programs the participants had taken the SM scale, so it was possible to see whether various markers of career success were linked to self-monitoring orientation. The results showed a clear pattern of career progression for HSMs. Relative to LSMs, they were more likely to achieve promotions by changing employers and moving locations over the five-year period. Even for those who stayed with the same company, HSMs received more promotions.

One of the subtle ways in which HSMs advance the likelihood of promotion is that they present themselves in their jobs to show their fitness for the *next* level of management to which they aspire. LSMs, in contrast, show more commitment to their organizations and are less likely to craft images of themselves that are conducive to promotion. There are potential downsides to each of these self-presentational strategies. LSMs run the risk of being seen as rather artless in terms of the image the group is trying to project.[15] Everyone loves Chuck in Tech Support, and his Twisted Sister T-shirt is classic Chuck, but it may draw censure when the negotiations that afternoon require him to wear a suit and be Charles with a button-down client. HSMs also run a risk by being too obvious in their acting above their level. Their peers in particular might see them as presumptuous and ostentatious.

In fact, there is evidence that HSMs do not appreciate or value performance appraisal schemes that involve peer-evaluations. They prefer to be evaluated by their bosses.

The stance that HSMs take toward their organizations is similar to that which they display toward their romantic partners—flexible but noncommittal. Whereas LSMs are more likely to form a few strong bonds of friendship within their work group, HSMs attend more to the broader network of group members. Within these networks HSMs assume central connecting roles, linking people who otherwise would be unlikely to be connected with one another.

SELF-MONITORING ABILITY: DO YOU KNOW *HOW* TO ACT LIKE THAT?

Although the typical assumption in the research on self-monitoring is that LSMs are not *disposed* to acting against their settled style of behavior, is it possible that they actually may lack the ability to act in such a fashion? The notion of personality as an ability rather than disposition has had limited attention within academic psychology, but research exploring that possibility is highly instructive. Participants were fraternity members who were asked to view two of the famous Thematic Apperception Test (TAT) cards and to explain "what was happening in the scene." One of the cards is known to elicit a moderate level of hostility themes, the other moderate levels for sexual responses. Immediately after they had completed their own story they then participated in a procedure known as "testing the limits," in which they were explicitly asked to write stories with particular themes—specifically, write the most hostile story you can and write the sexiest story you can. Here are two different fraternity brothers' responses to the request to write "the sexiest story of which they were capable."[16]

> *Story 1.* Martin leaned over her shoulder, and though appearing to look at her face, he was actually viewing the bulge at her chest. He was suddenly overcome with desire, wanting to grab them, rip open her

dress, put his mouth all over them, suck them, bite them—and lower, yes lower. His hands wanted to roam, to enter secret places, to weave through pubic hairs. But his mouth and tongue too were moist with the desire to lick.

Suddenly he seized, and she let out a moan deep in the throat—half-passionate, half-terrified. But she let him anyway; she wanted it. She quickly unbuttoned and his hands wormed into the opening. He pulled her to him, waiting and trying to touch all over as her dress fell. "Do you want me to touch you too?" she pleaded. Yes, yes! Here? Yes! And here? Yes! And here? Yes, yes!

Now compare that to this example of "sexy" writing:

Story 2. The young woman has been living with another man for several weeks. Now she is pregnant and has come to her father for help. The old man is at first shocked since he thought his daughter was away at school. But he advises his daughter not to marry her lover if he does not really love her as she says. He is not a prudish man and understands that these things can happen. He sees nothing wrong in keeping the child when it is born even though the mother is unmarried.

In the expectation that you've read enough, I've suppressed the last couple of paragraphs. But I think it is pretty obvious that these two individuals, independent of their writing talent or moral dispositions, had very different capacities to generate steamy narratives. The writer of story one, apart from having read too much James Joyce, has no difficulties whatsoever in engaging with the domain of erotica. The writer of story two, despite being given instructions to be as sexy as he could be, wrote something that was unlikely to provoke thoughts of unbridled passion.

So the question this leaves us with is a possible ambiguity in depicting high and low self-monitors. High self-monitors may have both the disposition and the ability to shift their self-expressions to fit the situation they are in. Low self-monitors may have neither the disposition nor the ability to make those shifts. There is very little empirical work to draw on to help us make those distinctions. But I believe it is important to remember that the LSM who appears to be uninterested in presenting a different self to different audiences may actually wish to do so but simply does not have the skill. Alternatively, he might well have the same capacities as an HSM and once even displayed it for all to see after too many single malt scotches in South Beach, but back in his regular haunts he just isn't disposed to be anything other than himself.

Although little research has been done on this topic with high and low self-monitors, one highly relevant study has been reported that looked at how self-monitoring orientation predicted performance skills. Participants were asked to act out spontaneous comedic sketches in homogeneous groups of HSMs and LSMs. The HSMs were better able to do this than were LSMs according to their own evaluations and, more importantly, those of independent judges. It appears then that HSMs may not only be better chameleons but better *stand-up* chameleons as well.[17]

SELF-MONITORING PRESS: UNDER THE CIRCUMSTANCES

Recall that at the time Mark Snyder developed the concept of self-monitoring, the field of personality psychology was embroiled in a debate about whether it was people's traits or the situations into which they entered that had a greater impact on daily behavior. Snyder's creative resolution was to postulate that there was a stable trait of personality that predicted in which direction individuals would tilt, and we have spent the chapter illustrating some of these important differences. But, ironically, this resolution left unexamined the question of whether situations themselves can create pressures in which most people, LSMs and HSMs alike, would be strongly disposed to act in a particular way.

One way of examining this is to invoke the concept of *situational press*. One of the founders of personality psychology, Henry Murray, developed this concept to explain the strong normative pressures that arise in different contexts.[18] He made the case that for every human need, such as those for affiliation or achievement, there is a corresponding environmental press (the plural is also press) that could facilitate the expression of that need. Environments rich in possibilities for social interaction provide the right press for those high in affiliation. Firms that encourage unrelenting competition and hold retreats at which you play paintball really hard provide the appropriate press for those high in achievement needs—and possibly for those with masochistic needs.

But might it be possible to scale situations or settings not just in terms of whether they provide appropriate press for needs and traits but also in terms of self-monitoring itself? In other words, might environments differ in terms of whether they call upon people to monitor what they are doing very carefully, whatever that might be, or whether they are those in which it is fine to do whatever you feel comfortable doing? I ran just such an exploratory study with two of my undergraduate classes.[19]

First, one group of students was asked to list a diversity of situations, places, or settings students might face during their years at university. They were not constrained to select only settings that were frequently encountered, although these were the ones that appeared most often. After screening out redundant or highly similar settings we then compiled a list of forty that were then administered to another group of students. These students were asked to look at each situation and determine its self-monitoring press—that is, to what extent it was the kind of situation in which people should be monitoring their conduct very closely.

The *highest* in self-monitoring press (SM-press) were:

1. Job interview
2. Public speaking
3. Appearing in court
4. Meeting the dean of the university

5. Funeral
6. Giving a seminar in class
7. Serving customers
8. First date

I have talked to several deans who were not pleased to see themselves lodged between a court appearance and a funeral, but I hastened to tell them that I believe the student raters construed the appearance to be one regarding a disciplinary matter. In those, deans can be deadly. One common feature of most of these situations is that they are occasions when others will judge the participant and that judgment will have personal consequences.

The funeral example is fascinating because, at first glance, it does not seem to provide as much immediate risk of judgment and censorship as the others do. But it was precisely at a funeral that one of the great comedic scenes in television history took place. Mary Tyler Moore, an HSM if ever there was one, was outraged that her coworkers were joking about the death of Chuckles the Clown who, as marshal of a circus parade, had dressed up as Peter Peanut and had been killed when a rogue elephant tried to shell him. But at the funeral Mary found herself stifling repeated laughs, whereas the others showed appropriate funereal demeanor. The scene reaches its climax as the minister explains to Mary that it was all right to laugh because that is what Chuckles would have wanted. At which point, of course, Mary starts wailing inconsolably, the ultimate humiliation for a high self-monitor!

The following situations or settings were rated the *lowest* in SM-press.

1. Sick at home
2. Watching TV with friends
3. Rock concert
4. Camping alone
5. Talking to a close friend
6. At the beach

7. Grocery shopping
8. Dinner at McDonald's

These situations are largely informal, and it is unlikely one will be judged or evaluated when in those settings. Some, such as camping alone and being sick at home, are by definition done in isolated settings in which you can display indiscretions and ignominious bodily noises with utter impunity. Others show how hanging out with friends can modulate the need to self-monitor. The results for McDonald's are noteworthy because one of the highest SM-press situations was "first date." These results would suggest that if you want to minimize SM-press, consider heading off to McDonald's rather than Domaine de Châteauvieu for your first date. That way you can be lovin' it together without worrying about inadvertently slurping out of the finger bowl. Of course, it is the HSM person who would in fact worry about such matters. Some of you might already have Googled the proper use of finger bowls and the menu for Domaine de Châteauvieu.

PRINCIPLED VS. PRAGMATIC? CONTRASTING VALUES IN SELF-MONITORING

Whenever I discussed self-monitoring with my students, questions of central importance to them came tumbling out, and class discussions could get pretty heated. I found out from some of my students that one couple in the class had actually severed their relationship after discussing their very different scores on the SM scale, although I suspect there were other factors leading to the severance. But one thing seemed clear: self-monitoring raises issues that get us directly into the weighty areas of morality, ethics, and values.

Snyder and his colleagues have proposed that LSMs adopt a *principled* way of acting and interacting with others, whereas HSMs adopt a more *pragmatic* approach.[20] Like philosophical Kantians, LSMs stick to their core convictions even when it might benefit them to do otherwise. Principles matter for them. Categorically. HSMs are more utilitarian and pragmatic in the sense that they will click into action those aspects of themselves that are best suited to the demands of the situation, even if it means there may be inconsistencies in their behavior. But there are two issues I think are important to raise before we draw too strong a contrast between principled LSMs and pragmatic HSMs.

First, the contrast seems rather invidious. By posing the contrast between principled and pragmatic, it implies that HSMs are relatively unprincipled and not guided by core values that matter. But it is possible that what drives HSMs to behave as they do arises not through mere pragmatism but because of a commitment to a *different* principle—a valuing of sensitivity to others and of accommodation to something beyond oneself. The boyfriend who was nice to your younger brother may have seemed a bit disingenuous, true, but it is possible that his motive was not one of controlling the situation or manipulating others. Your kid brother is a straight shooter, and your boyfriend found it refreshing to talk to someone at the gathering who wasn't locked into an overly zealous spin cycle. In short, it is possible—even likely—that both LSM and HSM individuals are acting on the basis of principled values. LSMs are following the principle of consistency and forthrightness; HSMs are following the principle of care and connection.

The second reason why we need to be careful when appraising the values of high and low self-monitors is because both orientations, if taken to the extreme, can be unprincipled and imprudent. They might even be pathological. Take the case of someone who is extremely low on self-monitoring. Being unwilling or unable to adapt to the situational demands of daily life seems unduly rigid and potentially maladaptive. In a simple world of black and white and good and bad such an orientation might be adaptive, but in a world of shades of gray and constant flux LSMs may find themselves challenged and unable to make progress on life tasks that require some flexibility. Despite their admirable constancy, they may find themselves at times utterly beleaguered.

Might extremely high self-monitors also be at risk for pathology? The term *aesthetic character disorder* has been invoked to characterize individuals who are fully engaged in the aesthetic moment with whomever they meet. But they can then rapidly turn to another person or engage a different project in which diametrically opposed values are expressed. This trumping of the ethical by the aesthetic, could be seen as HSM taken to the extreme. It might be suggested that on this spectrum one can find a full range of socially astute but manipulative people, from beguiling serial charmers oblivious to the hurt they cause to full blown psychopaths.[21]

ADAPTIVE FLEXIBILITY: SELF-MONITORING RECONSIDERED

So what have we learned from the study of self-monitoring that can enrich our reflections about personality and well-being? Like the Big Five traits that we discussed in Chapter 2, research on self-monitoring gives us insight into our friendships and intimate relationships as well as our likelihood of vocational success.

And once again we have seen that well-being is both complex and contestable. It is complex because being high or low on dispositions like self-monitoring will promote some aspects of well-being but simultaneously undermine other aspects. The advantages of being an HSM include a suppleness and agility that increases success in getting along and getting ahead in life. But the downside of being an HSM is that it can foster a lack of commitment to partners and organizations, a sense of self-fragmentation, and a reputation for being all things to all people. The opposite pattern, of course, holds for the LSMs: their constancy and commitment can lead to enduring relationships, but they may find that their ability to accommodate to changing contexts is frustrated, and because of this, their capacity to succeed becomes compromised.

The contestability of well-being arises from the fact that people may disagree strongly about which of the aspects of well-being are worth achieving—that is, about what we *ought* to do. HSMs likely regard flexibility to be something not only desirable to achieve but also worthy of achieving. LSMs, in contrast, might regard flexibility as a less worthy aspiration than constancy and forthrightness. Knowing where you stand on this dimension of personality helps you understand which of the various facets of well-being you are likely to experience. But a deeper awareness of self-monitoring also helps you clarify whether these facets are worthwhile, whether they have value in your life.

The research on self-monitoring gives us a valuable perspective on how HSMs and LSMs give differential weight to traits and situations in guiding their daily behavior. But I want to close by suggesting a somewhat different way of looking at self-monitoring. I think we need to reconsider the assumption that self-monitoring is primarily trait-like. I believe that the simple contrast between high and low self-monitors obscures the possibility that *both* of these modes of self-expression may be active in each of us and that we should look at *flexible* self-monitoring as the most effective stance to take when pursuing the projects and solving the tasks of our daily lives.

Imagine that you have the following schedule of events for next Wednesday. You have three important meetings at work, an evening at home where you really want to relax and spend time with your family, a Skype call from your best friend who really needs a best friend right now, and a visit to the vet to complain about being charged for your male cat's hysterectomy. If you are an HSM, it is possible that you would approach each of these as situations requiring a different aspect of yourself to be presented—a sharp and competitive you at work, a fun and loopy you at home, a patient and sensitive you on Skype, and a calm but forceful you when you confront the confused vet. LSMs facing the same midweek schedule would, according to self-monitoring theory, be unlikely to present a different self in these different situations. Although there may be a *tendency* for individuals to tilt one way or another as predicted by self-monitoring theory, I think *most* people, even the LSMs, would likely show flexibility in how they would present themselves in different situations this coming Wednesday. I think most people would shift from more formal to informal self-presentation as they move from the business meeting to the family room. In other words, I think people are very much aware of SM-press and when they have to respect situational norms.

I think self-monitoring needs to be appraised in terms of how it facilitates or frustrates adaptive functioning. Either high or low self-monitoring can be adaptive if the situations and contexts require it. For example, high self-monitoring is adaptive if you are living in a differentiated environment that requires a diverse set of selves to be presented. Modern urban living is like that, in contrast with traditional rural living. In those more traditional communities being an HSM was probably not adaptive—you might be thought duplicitous or, because you were unpredictable, shifty and a potential troublemaker. In such an environment low self-monitoring would be adaptive. (In Chapter 8 we will turn to a deeper discussion of how the places we inhabit help shape our personalities and advance or restrict our well-being.)

If you are still puzzling over your own self-monitoring status and you wrote the letter Q on your forehead, here is what the results indicate. Those who put the tail to right side as viewed from inside your head looking outward are more likely to be LSMs, and those with the tail to the left, HSMs. The logic is that HSMs convey the information literally from the point of view of the audience, whereas LSMs do it from their own perspective. However, I urge you to take this demonstration with a few grains of salt before you devour it whole. Of course the HSMs will have already sampled it and salted to taste.

chapter five

Control, Agency, and the Shape of a Life

I want to live so that my life cannot be
ruined by a single phone call.
FEDERICO FELLINI, *La Dolce Vita*, 1960

Reality cannot be ignored except at a price; and the
longer the ignorance is persisted in, the higher and the
more terrible becomes the price that must be paid.
ALDOUS HUXLEY, "Religion and Time," 1949

Now I believe I can hear the philosophers protesting
that it can only be misery to live in folly, illusion,
deception and ignorance, but it isn't—it's human.
ERASMUS, *Praise of Folly*

I F THE INTERACTION OF OUR INNER PERSONALITY AND THE
outer reality of the situations we find ourselves in, brought together
by the projects in which we engage, shape our behavior and lives, two
important questions arise: Do our own actions or forces beyond our
control ultimately determine our fates? Does having or believing we

have control matter? Are we agents who shape our lives, or are we passive recipients of whatever forces might play upon us?

The question of how much control we have over our lives has inspired conflicting answers for millennia and is still being hotly debated. Psychologists are contributing important insights into the technical and philosophical aspects of the question. However, personality psychology examines something different—the impact of our personal *beliefs* about control on our well-being. Some of us approach life with the firm conviction that we are the controlling agents, with luck or chance playing a very minor role. Others believe, with equal conviction, that forces external to us determine what happens in our lives, for good or ill.

To help you evaluate your own stance on this important question complete the following scale.

SPHERES OF CONTROL[1]

Write a number from 1 to 7 to indicate how much you agree with each statement.

Disagree			Neutral			Agree
1	2	3	4	5	6	7

_____ 1. I can usually achieve what I want if I work hard for it.

_____ 2. Once I make plans, I am almost certain to make them work.

_____ 3. I prefer games involving some luck over games that require pure skill.

_____ 4. I can learn almost anything if I set my mind to it.

_____ 5. My major accomplishments are entirely due to my hard work and ability.

_____ 6. I usually do not set goals because I have a hard time following through on them.

_____ 7. Bad luck has sometimes prevented me from achieving things.

_____ 8. Almost anything is possible for me if I really want it.

_____ 9. Most of what happens in my career is beyond my
control.

_____ 10. I find it pointless to keep working on something
that's too difficult for me.

INTERPRETING YOUR SCORE: Take the sum of your scores on items 1, 2, 4,
5, and 8 and add 35 to them. Then subtract the sum of your scores on
items 3, 6, 7, 9 and 10. This number is your score on Personal Control.
Based on young adult norms scores of 60 or more are regarded as high internal, and scores of 48 and less are regarded as low internal (or external).

The early research on this dimension of personality referred to it
as an internal locus of control versus an external locus. I'll shorten
the terms here to an *internal* versus an *external* orientation or, more
simply, *internals* vs. *externals*. Despite the similarity of labels, this dimension is not the same as introversion-extraversion, nor is it strongly
related to the other Big Five dimensions of personality we have already
discussed. Although the disposition to see our lives as under our control is sufficiently stable to be regarded as a trait, our experiences can
change it in an enduring way.

Internal orientation has been shown across many studies to have
a major positive impact on human well-being and accomplishment.[2]
Consider the following four areas in which high internals seem to be at
an advantage relative to high externals.

Resistance to Social Influence

One of the early classic studies in social psychology demonstrated the
power of social influence on perception.[3] Imagine you are a participant
in an experiment on how accurately you can make perceptual discriminations. You are in a room with five other participants. You are asked
to judge whether two lines flashed briefly on a screen are or are not the
same length, a fairly straightforward perceptual task for which there
is a clearly correct answer. As the experimenter asks each of the group
in turn, you hear the others say, "Same, Same, Same, Same, Same,"

and now it is your turn. What you don't know is that all the others in the groups are confederates of the experimenter, and they have been scripted to give incorrect answers. What would you do when you see two lines that are not equal in length but everyone else says they are the same? There is a strong tendency to be influenced by the group consensus. Even though the lines were clearly different, the pressure to misperceive is very strong. In the actual experiment this is exactly what happened: the real participants conformed with the consensus opinion, indicating that they thought the lines were the same. In short, people were willing to be influenced in ways that made their judgments suffer. However, subsequent research showed that there was a group of individuals who were relatively *resistant* to influence—those who scored high on a measure of internality.[4] Internals may have been puzzled that other people saw the lines differently from how they did, but they did not hesitate to declare their own judgment. Externals, confronted with the same social pressure, were the most likely to yield to the majority decision.

Attempts to get people to change their attitudes show a similar pattern. In one study you are asked to evaluate a new system for grading courses before and after you hear one of two short speeches that endorsed the new system. In one condition you hear a factual but low-keyed speech; in the other you receive a much harder pitch, being strongly admonished that you would be stupid not to vote to change to the new system. Would you change your view? The externals changed their attitude in response to both speeches, a little bit for the low-key and even more by the hard sell. But neither of the attempts moved the internals. In the more moderate influence condition they didn't budge at all. More strikingly, in the more extreme influence condition internals actually moved in a direction *opposite* of that being advocated. Some of us call this the POY ("Piss On You") response—a clear warning for people who push internals too hard![5]

Maybe internal individuals are simply rigid and become defiant when they are pushed to change their minds. But the evidence suggests that this is not the case. For example, whereas the prestige of someone attempting to influence them is more likely to influence externals, internals are more sensitive to the content of a message and will change if a compelling case is made to them.

Imagine trying to get an internal versus an external to stop smoking. In an intriguing study at Yale, participants, who were all smokers, were asked to play the role of someone who is given a diagnosis of lung cancer, complete with X-rays of the damage. They also took a measure of internality. The internals were more likely to actually decrease or stop smoking entirely after the experiment, but the externals were unaffected. It appears that internals will indeed change, but only if they have been convinced by logic or have personal experience, even simulated experience, that brings the issue clearly into view for them.[6]

Conversely, externals appear to be more fatalistic regarding things like illness and accidents, believing that sheer luck or good fortune are more likely to play a role in their health and well-being. This more fatalistic orientation of externals was nicely illustrated after I had given a public lecture on this topic and told the audience about the smoking study. A ruddy-faced middle-aged man with a big grin came up afterward and said that he completely understood why externals were unlikely to change their smoking habit. He told me that one reason he still smoked was because his grandfather smoked two packs a day and died at ninety-one from a massive orgasm. He also told me that he personally had scored as an extreme external and was proud of it. Like his grandfather before him, he looked forward to staying lucky right up to his dying breath. For this fatalistic external, life would conclude not with a whimper but a bang.

Risk Taking

The smoking study raises the question of whether internals are more likely in general to avoid taking risks than are externals, and research confirms that this is indeed true. When drivers in Detroit were stopped at a light and asked a few questions to tap into their level of internality, those scoring high were observed to be wearing their seatbelts more than those who scored low (admittedly, the study was done before cars were made to buzz incessantly when seatbelts are not buckled).[7] Similarly, highly internal university students were more likely to avail themselves of contraceptives than were externals.[8] If we consider the personality characteristics from Chapter 2, it would seem quite likely that a person having three characteristics—high extraversion, low conscientiousness, and an external locus of control—might also be at risk for a fourth characteristic: being pregnant.

Have you ever stood in line behind someone at a checkout counter who is choosing lottery numbers? And you waited, and waited, shifted your feet noisily, and found yourself actively suppressing the desire to shout, "Choose anything, you IDIOT"? There is a good chance the person in front of you has an external locus of control and an equally good chance that you are an internal. Externals invest themselves more fully in chance events than do internals; that is, they prepare for them and engage in the task with greater intensity.

But with skilled events the reverse is true. For example, consider which of the following approaches you would take if you were entered into a basketball free throw contest. You would get three points for a standard free throw but ten points for a shot that was truly from "downtown"—beyond the three-point line and twice the distance of the usual throw. If you had to choose just one of these positions for a two-minute session in which those with the greatest number of total points scored in that period were the winners, which would you choose? Based on research with a simulation of just this kind of condition, the evidence again shows that internals are less risk taking (shooting frequently from closer in), whereas externals are more willing to take a chance on the less probable course of action (long, risky shots).

My interest in locus of control and how people perform in demanding situations has resulted in some strange encounters. One resulted in me being accused of putting a hex on Wayne Gretzky! Here's how it happened. In 1980 I gave a talk at a communications conference on stress and control, which was then a very hot topic. In the question session afterward we started to talk about how some people, confronted with a challenge, may freeze and play well below their skill level. I suggested that this might be more likely with internals, who invest themselves heavily in skilled tasks, in contrast with externals. A fellow who said he had lived down the street from the Gretzky family in Brantford, Ontario, when growing up said that he figured Wayne would be an extreme internal because of his practice regimen and extraordinary work ethic. This sounded right. I then predicted that Gretzky might well have difficulties with penalty shots, where there is considerable pressure both in terms of personal aspiration and external expectation. Sure enough, in the next game Gretzky played, the first time he had a penalty shot, he missed. Next time, also a miss. Then he missed on his second penalty shot. The same on his third. Finally, after he scored on his fourth penalty shot, I got a phone call from an Edmonton radio station asking me whether I had taken the "hex" off Gretzky, and if I hadn't, could I explain why my prediction had been disconfirmed on the fourth try. The only thing I could think to say was that Richard Brodeur, the Vancouver Canucks' goalie who let in the goal, probably had an even higher internal locus of control score than Gretzky!

Linking Ends to Means: The Proactive Personality

The evidence so far is that internal individuals stand up to the influence of others without being rigid and are more invested in skilled performance than in performance that is based on chance or luck. An even more consequential difference is that internals are more likely to adopt a proactive approach to the projects and goals they are pursuing, and externals are more likely to adopt a reactive approach. By proactive, here, I mean the tendency to plan ahead and know how to link your aspirations to specific means that will accomplish them. One of the earliest confirming examples of this proactive orientation of internals was the data reported in the Coleman Report on factors that promoted academic success in American schools.[9] The best predictor of successful achievement was not a factor that might have been anticipated, such as intelligence or socioeconomic status, but rather a short measure of internal locus of control. Internals experienced more success than did externals. Although this finding was controversial, there is now increasing evidence, particularly from economists, that an internal locus of control positions adolescents for more positive trajectories both in education and in the workforce.[10]

Another example of how internals are more proactive was discovered in research on prisoners who were eligible for parole and who had completed a locus of control test. Internals were more likely than were externals to know how the system worked, when to apply for parole, and how to present the case to the warden. As a result, they were more likely to obtain their "get out of jail" card in a timely and effective manner.[11]

Delay of Gratification: From Marshmallows to SATs

One reason that internals are better able to shape their lives in productive ways is that they are better able to delay gratification—to "wait for it." In a series of highly influential studies Walter Mischel and his colleagues studied the capacity to delay gratification in four-year-olds.[12] The children were invited into a testing room, one at a time, and then told that the experimenter had to leave. A marshmallow sat on the table in front of the children. The experimenter told the children that they could have one marshmallow straight away or they could wait until she got back, without specifying how long that would be, at which point they would get an extra marshmallow. The children were monitored to see what they would do. Some children wolfed down the marshmallow right away, some hesitated, and others waited until the experimenter came back and were then able to claim two of them.

It is instructive—and often delightfully amusing—to watch the way these little four-year-olds dealt with the task of delaying their gratification. Some put their noses right on the marshmallow in front of them, and others looked away and engaged in various forms of distraction. Those who had the distracting skills down pat were the least likely to give in to the temptation. The intriguing aspect of this set of studies was the follow-up research that looked at these same individuals many years later. Those who had been able to resist temptation and wait for the second marshmallow had better school performance and performed significantly higher on the SAT, a demanding test college and university selection committees use.[13]

STRESS, CONTROL, AND BUTTONS

Imagine you are a participant in an experiment on noise stress. You are brought into a research lab and told you are going to perform a simple clerical task while listening to blasts of loud noise through your headphones. You don't know when the blasts are going to come, and the noise, though not dangerous, is highly unpleasant—about the same as you would experience if you stood near a jet plane's engines (in fact, the noise was a recording of jet engines). You perform the task, and as you do it you are monitored for your level of autonomic nervous system arousal (such as blood pressure, heart rate, and sweating). After you finish that phase of the experiment you are put in a rather crowded room with other participants, where you are asked to complete some tasks, some of which seem virtually unsolvable. How do you perform on the last phase of the experiment? More specifically, how do you do in comparison with other participants who don't experience the noise or others who *can* predict when the noise will occur?

These were the essential questions driving an important set of studies carried out at Rockefeller University.[14] The overarching question was whether people would be able to adapt to the noise stressor so their performance in the subsequent task wasn't compromised. The results were clear and compelling: although at the beginning of the noise interruptions the participants showed an increase in autonomic arousal, after a short time they adapted to the noise and their arousal returned to normal. However, in the subsequent task they exhibited some *costs* of adaptation. Those exposed to the noise, in comparison to a control group, made more errors and exhibited more signs of frustration and hostility in their subsequent performance. Given this result, what might be the effect of having a sense of control over the source of the noise stressor?

Two variations on the experiment procedure provided some highly instructive answers. First, it was shown that participants who had experienced random noise bursts adapted less quickly than did those who were exposed to predictable bursts. This could be seen as a type of control—not direct control but an anticipatory one that made the subjective stress more tolerable.

Second, and crucially, was a variation of the experiment in which individuals were told that they actually had control over the noise if it became too punitive for them: they could press a button that would stop the noise. In published studies it was noted that very seldom was the button actually pushed, and eventually the button wasn't actually hooked up at all because it was never used. However, the group in the button condition who had a sense of control showed notable benefits compared to those who did not: they adapted physiologically to the noise, returning to their base rate arousal level more quickly, and they also paid less of a cost for this adaptation, making fewer errors and displaying less frustration and hostility in the subsequent testing session.

I found these results intriguing, particularly in their applicability to the stressors of everyday life. Consider a typical morning commute. If you are driving, there is relatively little you can do to anticipate a clear and unimpeded route to your destination. It is likely that as you begin your commute you experience a sense of physiological arousal, but this would subside as you begin to adapt to the morning routine. But even though you may adapt *physiologically* to the commute, you might be expected to pay a *psychological* cost when you finish the drive. When you get to the office you are more prone to make mistakes and more short-tempered than if you had not had to adapt to the stressful commute. But believing that at any point you could bolt from the highway and take an arterial road to your destination should reduce that stress and lower the costs, as would getting up at 4:30 a.m. to ensure an unimpeded route to your destination.

I don't think it is too much of a stretch to see the button condition in the studies of stress and adaptation as the equivalent of an internal sense of control. Those high in internality could be seen as having a bunch of buttons they can push as needed to cope with the daily stressors of living. Confronted by a demanding exam? Push your "study hard" button. Fascinated by a potential romantic partner? Push your "charm" button. Facing an uncertain future? Push your "optimism" button.

Given the evidence I've presented so far, I'd imagine most readers would now conclude that an internal locus of control or, more broadly, an "agentic" orientation (the sense that *I* am in charge) helps advance our aspirations in life. Let's summarize what we know: internals are more likely to resist unwanted influence, avoid undue risk, and make clear plans to achieve their valued goals. They are able to delay short-term rewards for larger, more distant rewards and are better able to deal with the stresses of everyday life as well as pay less of a cost for exposure to them. It is hard to question the benefits of such an orientation to life. . . . Or is it?

Are Your Buttons Hooked Up?

Some years ago I was a presenter on a panel at a multidisciplinary conference on Stress and Resistance to Change. I focused on locus of control in everyday personal projects and covered many of the studies discussed in this chapter. At the end of the presentations there was a Q&A session, and a fellow sitting by the door at the far back had a question for me: "Professor Little, you mentioned that in some of the studies the button wasn't actually hooked up. Is that right?" I answered that, yes, I had seen a footnote to that effect in one of the studies in that research program. "What do you think might have happened if someone pressed the button and found it was not hooked up—wouldn't they be even more stressed than those who never thought they had control in the first place?"

I thought this was a splendid question. I wanted to know what had inspired him to ask it and inquired whether he was a clinician who had dealt with issues of stress and control in his practice. "No, I'm not any kind of psychologist. I'm a political scientist, and I'm actually in the wrong room, but when I realized it was too late to leave graciously I decided to stay. The reason the button condition intrigues me is that it perfectly models my theory of the relationship between governments and the individual. The government creates the illusion of control, individuals buy into it, and when they find nothing is hooked up they turn on the government."

The audience laughed, I laughed, and that seemed to be the end of it. But the question gnawed away at me, and for the next few months I kept my eyes open for anything that might have provided a convincing answer. Framed at a more abstract level, it dealt with issues such as the loss of control, the illusion of control, and whether such illusions were adaptive or maladaptive. I wasn't able to find a study that precisely examined this question, but I did find some articles in closely related fields that provided a route to a better understanding of control, illusion, and the shape of a human life. And simultaneously with this searching I had the chance to reflect on some personal concerns that raised the same issues, and in dramatic form.

The Realities of Perceived Control

At first the studies I unearthed confirmed the by now well-established finding that having a sense of control over events in one's life paid psychological and health dividends. One of the areas in which this was particularly well documented was in the field of gerontology and the effects on residents in nursing homes when they have control over everyday events. One of the downsides of entering a nursing home can be the loss of freedom and control that accompanies the transition from one's own home to a care facility. This loss and attempts to mitigate it formed the basis of several key studies. One experimental study by Ellen Langer and Judith Rodin in 1976 showed that making minimal changes to increase the level of personal control given to residents had substantial benefits. The changes involved were fairly simple: giving the residents more control over such things as what movies to attend, room decoration, and care of house plants. Compared to residents for whom staff controlled these aspects of daily life, those with personal control and responsibility increased their activity level, were happier and healthier, and even lived longer.[15]

At about the same time as Langer and Rodin's study, Richard Schulz and his collaborators carried out some fascinating research that had unanticipated effects.[16] The studies began with Schulz's doctoral dissertation at Duke University in which he explored how a sense of control over a valued activity had an impact on nursing home residents. The valued activity was social engagement. In the best facilities there is an optimal level of social stimulation for the residents, but in many there are residents who sit in ominous silence for much of the day. They ache for social stimulation. Schulz arranged for groups of Duke students to visit the residents in this facility under one of two conditions—either the residents had control over the visits or the students did. There was also a group who received no visits. The length of the visits and other features of the social engagement were the same in both experimental conditions, but, as expected, those residents who had control over the visits ended up having greater mobility, subjective well-being, and health than did those who did not have a sense of personal control. So far the studies had confirmed the expected relation between control and beneficial outcomes. Nothing new here, apparently.

But in fact there was something very new to consider. The study ended, the students graduated, and control of the visits was lost, abruptly and apparently without clear explanation. In a remarkable follow-up investigation Schulz revisited the groups that had been studied earlier. The group who had previously experienced control over the students' visits, relative to the group who never did have control, showed a marked reduction in health and happiness. And even more remarkably, their mortality rate was significantly higher. As I read over the results of the study I immediately thought of the question from the back of the hall about buttons and control. What happens if you have a sense of control over something you value but then you lose it—you press the button and it isn't hooked up? I thought to myself that such a loss of control in extreme cases might be, quite literally, deadly serious.

Class Acts

During the time period in which I was looking at the growing research literature I also had some experiences that alerted me to the personal side of losing a sense of control.

I have taught dozens of courses on personality psychology over the years, both at an introductory level and for more advanced students. Particularly when the size of my classes was manageable, say around thirty or forty students, I found it extremely useful to include a course-long exercise that I called the Personal Sketch.

On the first day of class I would give the instructions. Students were to choose a pseudonym and enter it at the top of their sketch, and then they were to write a two-page, single-spaced, double-sided essay that discussed their own personality. It could take any form they believed conveyed the essence of their personality—it could be a list of their most defining characteristics, an overview of their personal development from childhood, and so forth. Importantly, they were asked to write it in the third person and from the perspective of "someone who knew them extremely well, perhaps better than they knew themselves." When completed, each essay was copied and distributed to all other members of the class, so each member of a class of thirty would receive at the end of the first week of lectures a copy of twenty-nine other personal sketches.

The impact of sharing the anonymous personal sketches on the classes was remarkable. The undifferentiated blobs of humanity from the first couple of lectures became vivid and fully differentiated individuals. Except no one knew who was who, and I had warned them not to include any clearly observable physical characteristic that would compromise their anonymity. (I did have to delete an allusion in one sketch that described the writer as a member of the university basketball team, as there was only one person close to being seven feet tall in the class!) Throughout the course students would then apply readings and lectures to the further elaboration and understanding of their own sketches or those of other students, and these were submitted to me twice a term as a course "journal." Both the sketches themselves and the journal entries were fascinating. Three of them had a profound effect on me.

The first time I used this technique a woman who was about ten years older than the rest of the students began her sketch with "It has been asked whether we determine the course of our lives or if it is determined by forces beyond our control." It went on to describe her years, right up to her midtwenties, as filled with achievement, success, happiness, and a pervasive sense that she was in control of her life. Then, in one short paragraph, she described a series of events that shook her to the core—the death of her children in an accident, a betrayal, a divorce, a falling apart. She concluded her sketch by saying that she had muddled through these personal catastrophes, but they had stripped her of her natural *joie de vivre*. Clearly too she had shifted her orientation toward a view that the vicissitudes of life can deflect our lives in ways that are unfathomably cruel. She was moving ahead with her life, but she was doing so with a new awareness of some of life's fragilities. She could reflect on her vulnerability. She wanted to share this with her younger classmates, willingly relinquishing her anonymity. She was a class act.

A second experience with personal sketches led to me having my own brush with a major, unpredictable, and uncontrollable event. I was reading my students' journal entries halfway through term in which they commented on the relevance of course lectures and materials to their own lives. One journal entry had the heading, "Death Threat," and it was aimed at me personally. I can remember it word for word and particularly the last paragraph: "I am going to take the revolver my brother-in-law gave me for Christmas and I am going to shoot you right between your beady little eyes. Do not go near the river [that ran beside campus]. My gun is cocked."

I have to admit—and this is embarrassing to acknowledge—that my very first conscious reaction was to think, "What do you mean, 'beady little eyes'?" I was unable, until a few seconds later, to process that this could well be serious. The threat launched a series of surreal encounters. First, I needed to check to see whether this was a hoax or a credible threat. The head of my department, the dean, and, in particular, the psychiatrist we contacted at the university Health Services were convinced that this was not a hoax and was sufficiently serious to warrant calling the police. Because I knew the name of the student (his sketch was anonymous to the class but his journal was, of course, known to me), the police asked me to name him, which I refused to do. It was on a Friday, and he would have been arrested on a charge of issuing a death threat and held in jail over the weekend. But what if it were a hoax, albeit a rather inexplicable one? Also, the threat had arisen in the context of an exercise I had promised my students would be entirely confidential. There was considerable eyeball rolling and tongue clicking from the legal officers of the university when I said I would not press charges at that time. The campus security officers, normally not an overly busy crew, were most interested in the case—actually they were dead keen about it. They were worried about my office hours I was planning to hold on the following Tuesday morning and gave me detailed instructions on how we should deal with the situation. One of "their best men" (of three available) would be stationed in the room next to mine with a glass cupped to the wall, and if he heard any threatening noises coming from my room (perhaps a gunshot), he would "call the local police." This was not very reassuring.

I was, of course, terribly worried about the safety of my young children. But I found it difficult to get security at home without naming the person of interest whom police kept referring to as the "alleged student," a moniker that, to my beady little eyes, seemed technically incorrect. I was also concerned in case the students in my laboratory would be in danger from any confrontation. The security folks warned them about a possible threat, and there was an abrupt change in their behavior toward me. "Hey, Brian, could you get our letters of recommendation done as soon as possible, please?" I think they were kidding, but perhaps not!

The death threat turned out to be a hoax. The motivation behind the journal entry was never fully explained. Life returned to normal, and as far as I could tell, I did not experience any major undesirable side effects. What I did experience, however, was a change of orientation. Together with some other challenges I was experiencing at the time, the incident shifted my sense of control from a resolutely internal one to a sense of personal control that was more contingent and nuanced. Instead of *assuming* control, I learned to check my assumptions and scan my environment for potential dangers before they arose; in short, I learned to check my buttons very carefully.

About a month after the death threat scare a third incident involving a student and a personal sketch unfolded in one of my other classes. It involved a cigar box. In the previous lecture I had told the class about the experiments on stress and control and the button condition. I also told them about the question from the back of the room about what happens when you discover your buttons aren't hooked up. A tall, lanky, curly haired guy, as I recall an architecture student, came up after class and slipped me a note, saying that he was working on a journal entry about the topic. He added a note: "Check your office door." I was heading off campus but was curious, so I headed up to my office. There, hanging from the door, was a cigar box with wires coming out of it and a big brass button that had written on it "Press Me and See What Happens." Now I don't think of myself as a deeply stupid man, but it never really dawned on me that there was anything dangerous about this at all. I thought it was a clever way of symbolizing the lecture the students had recently heard, so I chuckled and then headed off.

The security staff, in their wisdom, remembered the events of the previous month, and from what I could tell from the residents of that corner of my building the next day, these custodians of campus safety did not chuckle. The following morning I went up the elevator to my office, and as soon as the door opened I saw my dean and the head of security waiting to intercept me. They proceeded to tell me that everything was okay now but that there had been an incident overnight at my office. Apparently security had seen the cigar box, button, and wires, and they had alerted the police. They came with a bomb squad and blew up my door. "Got her good," said the head of security. I'm not entirely sure I will ever understand the logic of demolishing my door, but the whole event seemed symbolically expressive of the surreal month I had just experienced. I was able to rescue the cigar box with its big brass button, and I have stored it away as one of my prized possessions. It reminds me of the potential perils of professing and the strange complexity of seemingly simple buttons—both real and metaphorical.

ADAPTIVE ILLUSIONS AND STRATEGIC SPIN

I imagine that the long list I've presented of positive consequences that go along with an internal orientation pleased readers who took the locus-of-control assessment at the beginning of this chapter and scored high on internality. This information is consistent with the advice that most of us have grown up with, especially those of us who have been well educated and encouraged to achieve at a high level. We have been raised to believe that we can control our destinies and that our boundaries are limited only by our imaginations.

But it is well worth contemplating those studies that sounded a warning about control beliefs that are based on illusions: the unhooked-up button in the noise stress studies, the unrealistic expectation of continued social visits in the retirement home research, and the poignant account of a woman whose life fell apart for reasons beyond her control as well as her testimony about the turmoil and pain that ensued.

We need to make sure that the buttons we think we have in our lives are, in fact, hooked up. To do so, we need objective feedback on our abilities in different domains and to face it unflinchingly. We also need to invest in projects that align with our talents and capabilities as well as our ambitions. And we need to scan our environments to see whether they help or hinder our pursuit of personal projects. For many, this is an uncomfortable and perhaps threatening process. But before committing ourselves to new ventures in our lives we need to be honest with ourselves and request honesty from those we turn to for advice and counsel. Our illusions are often the product of collusions with others who may or may not be invested in our long-term welfare. Consider this example of collusional illusions that I overheard in the university cafeteria one morning.

A professor from another department whom I had known only by his reputation as a tenured lecher was in line just ahead of me. Consistent with his reputation, he was accompanied by a highly attractive young woman with whom he was locked in mutual gaze. She was complaining about her courses. Although she thought English was fun, she hated math, was failing both physics and biology, and thought chemistry was "stupid." But she "really, really wanted to be a gynecologist." He looked deeply into her eyes and purred, "Go for it." I really, really wanted to yell at her "DON'T go for it. For the sake of the wombs of the women you want to treat, DON'T be a gynecologist. Write about wombs, create variations on the *Vagina Monologues*, but don't go into medicine!" But I didn't say a word. After all, who was I to challenge her illusions? I later found out that she failed all of her courses, including English, and had ended up in Oregon studying at some New Age college of wind chimes, tarot card reading, and exotic body massage. I suspect she would have done well.

But if she had been in my course and had attended the class on buttons and control, I could have had her at least reflect on her abilities and aspirations. Along with the rest of the class I would have respectfully challenged her to examine whether an illusory glow unduly influenced her life pursuits and core projects.[17] And I would have wanted her to make sure that those whom she entrusted with guidance on her life choices would be as objective as possible on her prospects, that they would have been willing to show her how some of her button connections were frayed and in danger of breaking. Although her fawning professorial friend was unable to give her the objective feedback she needed, her course grades did, and she ended up at last, so it appears, in an occupation in harmony with her personality, orientations, and abilities. She was clearly a delightful and engaging person who could well be flourishing today. But I suspect she might still need help with her buttons.

Perhaps the most central question the studies on internal and external locus of control orientation raised is whether some illusions might actually promote our well-being. There is much empirical evidence that people hold an abundance of *positive illusions*, such as the belief that we are in control of events that are objectively uncontrollable or that we have desirable personality characteristics when others say we do not. For instance, do you think you have an above average sense of humor? Virtually everybody does, which in terms of probability is unlikely to be true! Such illusions, as long as they are not too extreme, are actually adaptive and can enhance well-being.[18] And if we look at the other end of the well-being spectrum, at those individuals who are depressed, we find that they are *more* realistic in their perceptions of control and contingency than those who were not depressed. Might it be said then that depressed individuals, relative to the rest of us, are sadder but wiser? They are, most certainly, sadder. I also think that they might be more *knowledgeable*, if by that we mean having a more accurate read on the realities of perceived control and personal strengths. But wiser? I don't think so. It comes down to a matter of the timing of our illusions.

There are times when adopting positive illusions may be adaptive and times when it may sabotage the pursuit of what matters to us.[19] Knowing how to spin your projects to optimize the chance that they will be successful is a key aspect of wisdom. For example, when we are in the process of deciding whether to pursue a course of action, like committing to a field of study, changing jobs, or taking a relationship to a new level, it pays to tilt away from illusion and toward reality. Scanning broadly for relevant information about whether this is worth pursuing and whether it is likely to be a successful pursuit decreases the likelihood you will be stopped short in your pursuits. Failure to do so may lead to you being suddenly confronted with the unexpected—gobsmacked, as the Brits say.

Once we are committed to a project or pursuit, however, adopting a positive spin and not being distracted by the negative realities works to our advantage. It is in the heat of project pursuit that illusions become adaptive. But this is only after we have done a realistic appraisal of our own abilities, convictions, and the extent to which our everyday ecology will facilitate or frustrate such pursuit.

I began this chapter with three epigrams from three contrasting perspectives on control and well-being. Fellini's character in *La Dolce Vita* was committed to controlling his life so nothing could disturb his peace of mind. No phone call could challenge his well-being. Illusion was his guardian. Huxley's exhortation about the tyranny of illusion and the triumph of realism takes the opposite view: we abandon objectivity at our peril. The phone will ring, and we should unflinchingly answer it. Finally Erasmus reminds us that living with illusions is part of the human condition, perhaps a natural way of muddling through complex and perplexing lives. When we are young, as most of my students are, illusory beliefs in agency and control might well steer them in directions that prove frustrating or painful. But such illusions might also keep them in the game and allow them to spot new, more achievable possibilities for their lives. Despite our sometimes frayed button connections and happy illusions, there is one stance toward our lives that seems both circumspect and audacious. It is neither illusory nor constraining; it is both adaptive and contingent. It is something we need to incorporate into our reflections about how our lives have gone and how they will fare in the future. It is, quite simply, hope.

chapter six

Hale and Hardy:
Personality and Health

ONE BITTER FEBRUARY MORNING I WAS SITTING IN MY physician's waiting room about to have my annual health checkup when I witnessed something rather disturbing. The man sitting next to me—red cheeked, blue eyed, receding hairline, midthirties—was filling out what I assumed was a crossword puzzle in a magazine. But he seemed to be getting agitated. He suddenly blurted out, "Oh my God, I'm going to die!" My first thought was that he had chosen a pretty good place in which to realize this. I suppressed that thought. My second thought, which I didn't suppress, was to glance over at his magazine and see whether it might provide some clue about his sudden outburst. It did. It was a popular magazine that had a health questionnaire in it that I immediately recognized. What had he read and how reasonable was his concern?

If you promise not to scream out loud, you can take the questionnaire he had taken. Here it is:

Check off any of the following forty-three life events that have happened to you within the past twelve months.

Life Events Change Scale[1]

Life Event	Value	Check if this applies
1. Death of spouse	100	☐
2. Divorce	73	☐
3. Marital separation	65	☐
4. Jail term	63	☐
5. Death of close family member	63	☐
6. Personal injury or illness	53	☐
7. Marriage	50	☐
8. Fired at work	47	☐
9. Marital reconciliation	45	☐
10. Retirement	45	☐
11. Change in health of family member	44	☐
12. Pregnancy	40	☐
13. Sex difficulties	39	☐
14. Gain of new family member	39	☐
15. Business readjustment	39	☐
16. Change in financial state	38	☐
17. Death of close friend	37	☐
18. Change to a different line of work	36	☐
19. Change in number of arguments with spouse	35	☐
20. A large mortgage or loan	31	☐
21. Foreclosure of mortgage or loan	30	☐
22. Change in responsibilities at work	29	☐
23. Son or daughter leaving home	29	☐
24. Trouble with in-laws	29	☐
25. Outstanding personal achievement	28	☐
26. Spouse begins or stops work	26	☐
27. Begin or end school/college	26	☐

Life Event	Value	Check if this applies
28. Change in living conditions	25	☐
29. Revision of personal habits	24	☐
30. Trouble with boss	23	☐
31. Change in work hours or conditions	20	☐
32. Change in residence	20	☐
33. Change in school/college	20	☐
34. Change in recreation	19	☐
35. Change in church activities	19	☐
36. Change in social activities	18	☐
37. A moderate loan or mortgage	17	☐
38. Change in sleeping habits	16	☐
39. Change in number of family get-togethers	15	☐
40. Change in eating habits	15	☐
41. Vacation	13	☐
42. Christmas	12	☐
43. Minor violations of the law	11	☐
YOUR TOTAL _____		

Notice that each event is given a weight in the left-hand column. Add up your total score for the weights of the events you have checked off. Here are some comments on the health implications of your total score.

SCORE	COMMENT
300+	You have a high or very high risk of becoming ill in the near future.
150–299	You have a moderate to high chance of becoming ill in the near future.
<150	You have only a low to moderate chance of becoming ill in the near future.

LIFE EVENT CHANGES AND HEALTH

Now relax—despite its enormous popularity, the Holmes-Rahe scale, which you have just completed, has some flaws that I'll point out in a minute. But you need to know why the man next to me in the doctor's office—Chad, he told me later—had yelled out after adding up his score. He had scored 423 on the scale, and having read in the magazine that scores over 300 were regarded as indicating a "high or very high risk of becoming ill in the near future," he was mightily concerned.

I chatted with Chad for a few minutes and convinced him to be cautious about interpreting his score. We'll get to that discussion in a moment. But first, a few words about the limitations of the scale.

In the mid-1960s psychiatrists Thomas H. Holmes and Richard H. Rahe developed the scale, wishing to create a simple measure for use in epidemiological studies of stress and health problems. Their underlying rationale of the scale was that stress would result from interruptions to daily routines and that some life events would have a major disruptive influence. Stress, in turn, was expected to lead to compromised health and a diversity of medical problems, and some of the early research showed a significant, albeit extremely modest, correlation between life event change stress and health problems. One of the most interesting aspects of the scale was its assumption that positive events, like marriage or getting a new job, would contribute to stress because they would increase disruption in people's lives.

Like the Myers-Briggs Type Indicator we met in Chapter 2, the Holmes-Rahe scale figured prominently in the popular press, and people were intrigued by and, occasionally, like Chad, alarmed by their results. Despite having some real strengths, there were a number of problems with the scale, and this is one reason why I cautioned you to be careful about any conclusions you draw about your own health after having taken it.

First, consider the weights that were assigned to each of the events. They are based on researchers' estimates about how disruptive an event would be, based on the death of a spouse as a standard of one hundred points. But consider how these standardized weights could obscure the *personal* weights that individuals might give to these events. A woman who has watched her husband suffer for years might greet his death as a welcome respite from his pain. Notwithstanding her inevitable grief, she may find that her stress level was lower than that of someone for whom the death of a loved one left her completely anchorless and utterly inconsolable. Would it not make more sense to allow individuals to provide their own weights for the life events they had experienced? Second, the weights are *added* up. It is not unlikely that some events, such as moving cities after the death of a spouse, may *reduce*, not add to a person's overall stress level. Life events, in short, form systems, and we need to know how the various events relate to each other. Third, although including positive life events, such as marriage and a promotion at work, is of considerable theoretical interest, the research evidence shows that it is only the negative events, not positive ones, that predict subsequent health decline.[2] Fourth, notice that some of the events, such as difficulties with sex, eating, or sleeping, may, in themselves, constitute health issues. The prediction of health problems from prior health problems should not be surprising. It is like predicting that a tadpole has "a high or very high risk" for being a frog. The inclusion of such items can be, well, rather uninformative.

So what about Chad? We struck up a conversation, and he showed me his answers. We chuckled at the fact that by checking off "Christmas" he had already earned twelve points on his stress score. He had recently married and had moved back from where he had been completing a graduate degree to his hometown. His father had died six months earlier from a painful neurodegenerative disease, and Chad was glad to now live close to his mother who still lived in the family home. Today he also was in for his annual health checkup.

Notice how each of the shortcomings of the scale play out in our interpretation of Chad's likely health status. If we took into account only the negative events, discounting those that made other negative events less serious; if we ignored his marriage and move, both of which he described as wonderful; and if we discount his broken elbow, Chad's stress level and health risk were, I thought, pretty minimal. Although I didn't go into this much detail in the doctor's office, I did convey to him that he should be cautious when inferring the likelihood of health decline from these kinds of tests. Actually I just said, "Really, don't sweat it." But in retrospect I wish I had been able that day to give him more details. So Chad, if you happen to have picked up this book, as you read this chapter you will find out more about personality and health than I was able to tell you on that bitter February morning. And if you, like Chad, also had an all-too-eventful year, keep that scream suppressed and read on.

Personality, Stress, and Health: Hardiness and Resilience

Among the large number of people who have taken the Holmes-Rahe scale were employees at the Illinois Bell Telephone Company (IBT) in Chicago in the mid-1970s. Salvatore Maddi, a distinguished personality psychologist at the University of Chicago, with the encouragement and support of Carl Horn, one of IBT's executive vice presidents, began a long-term evaluation of personality, stress, coping, and health with a large number of IBT employees. Maddi and Horn had been aware of the inevitability of major disruptive changes that might occur at this facility due to divestiture and deregulation legislation affecting the telecommunications industry. In 1981 IBT was hit with a massive reduction in staff, from twenty-six thousand to about fourteen thousand, causing an inordinate increase in daily stressors. Throughout this period of turbulence the employees continued to be studied, providing an opportunity for the researchers to look in some detail at what happens to an employee's emotional and physical health during a period of major life event changes.[3]

The results of the study were intriguing. About two-thirds of the employees showed symptoms of health decline and lowered performance, but a third of the employees appeared to deal effectively with these changes and emerge resilient and unscathed. How did these two groups differ? The groups did *not* differ in terms of their scores on the Holmes-Rahe scale; in other words, those who survived the stresses of downscaling reported the same level of life event changes as those who did not. What did differentiate the two groups was a set of personality characteristics that Maddi and colleagues called *hardiness*. Hardiness comprises three key components: commitment, control, and challenge—the three Cs of hardy personalities. A sense of commitment was exemplified by an attitude of being fully engaged in everyday events rather than feeling isolated and excluded from them. A sense of control was displayed by employees who tried to exert influence over the life events changing around them rather than being passive and feeling powerless. A sense of challenge was an attitude toward change that led them to view both positive and negative changes as opportunities for growth and new learning. In short, this research and the extensive set of follow-up studies that it stimulated led to the following conclusion: *health is enhanced to the extent that control, commitment, and challenge are core aspects of an individual's personality.*

That is what I had wanted to convey to Chad. Even if he *did* have a valid high score on the Holmes-Rahe scale—which, as you know, I had doubts about—the experience of stress in everyday life is ubiquitous, and we shouldn't feel that the only healthy response is to avoid engaging with life, with all its attendant risks. Instead, Chad's orientation toward those changes, his way of coping with them, could mitigate most of the potential health hazards they posed.

TYPE A PERSONALITIES

Consider now a very different type of personality characteristic that also has implications for our health: the Type A, or coronary risk, personality. This is one of the most extensively studied concepts in the fields of behavioral medicine and health psychology. Most people have heard of it, and it has become part of our everyday conversation about people's lifestyles. When I talk to groups about Type A behavior I ask them to call out some of the characteristics they believe are associated with this style, and there is remarkable consensus about what it entails. The most frequently mentioned features of Type A people are time urgency, forcefulness, and competitiveness, and, as we will see, these do characterize what we might call the "surface" features of Type A personality. The following page from a personal calendar illustrates all too well some of the relevant characteristics of Type A behavior.

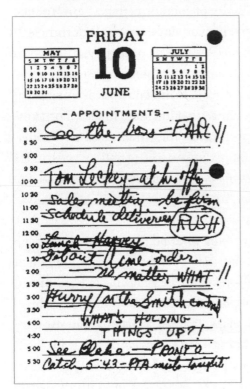

First, notice the clear evidence of time urgency. Things need to be done "EARLY," in a "RUSH," and "PRONTO" (all capitalized for emphasis). And notice the competitiveness with the self-admonishment to be firm and to achieve his sale target "no matter WHAT!!" In the final entry the writer exhorts himself to catch the 5:43 that night for a PTA meeting. As it turned out, that would be his last such meeting. Shortly after completing this entry in his calendar he died of a heart attack.[4]

The health problems associated with Type A personality are obvious. What might not be quite so obvious are some of the advantages that having such a personality might offer other aspects of our well-being, such as our level of accomplishment and occupational success. Think about the kind of advertisements we see for virtually any kind of job. They are seeking people who are hard working, very ambitious, highly committed, and who seek out challenges and who exhibit a take-charge attitude. I have not yet seen an advertisement that says, "Wanted! Unambitious, lazy people with an aversion to commitment." Mellow doesn't really cut it when it comes to applying for jobs. There is also evidence that Type As are more productive and can, therefore, reap the benefits of success, particularly in competitive environments. For example, in my own field of university teaching and research, Type A individuals have higher citation counts; that is, other scholars reference their work more frequently than that of their less-driven colleagues.

I should also mention that Type A individuals pose challenges not only for themselves but also for their coworkers, families, and friends. Their interaction style can be frustrating. They speak louder and gesture more expressively than do those who are not Type As. They are prone to interrupting others who, in their opinion, speak too slowly. They are blunt, impatient, and easily frustrated, and although this might actually advance their progress on things that matter to them, they may, in the process, find others avoiding them because interactions with them are so stressful.

One of the most interesting features of Type As' behavior is that they are insensitive to signals of stress in their own bodies because of the intensity of their focus on tasks at hand. I experienced this first-hand a few years ago. I was on a committee planning a major conference, and this entailed having a number of meetings over the course of the year. The person chairing the committee was clearly a Type A. His intensity was legendary, and when I would get off the phone with him, usually very late at night, I'd feel as if I should wipe my ear from spittle because of his explosive, punctate way of speaking. But it was in face-to-face meetings with the committee that his personality extracted its greatest toll. The committee typically met at 4 p.m. with the intention and expectation that it would review progress and wrap up around five at the latest. In one such meeting the chairman got agitated about a budget issue, and it was very clear to the other committee members that he was stressed out: he was grimacing, had a sweaty forehead, and was clenching his jaw and fists. But he didn't notice these signs of tension in himself, nor did he notice the obvious signs of fatigue and annoyance of the other committee members as seven o'clock rolled around. That nobody would say anything to him was a direct consequence of his deep-seated belief that this budget issue was the most important and consequential issue in the entire universe and that life as we knew it would end if we didn't get it solved. He was also a very large man with a very short temper. He wasn't elected to the planning committee for the next conference.

What, then, might we say about Type As that captures the gist of their personality but doesn't have too much of an evaluative element to it? I think a fair conclusion would be that there are three features that distinguish them. First, *control* matters to Type As, and they are particularly fearful of losing control once they have achieved it. Control means they are able to push through on their projects without delay. Second, Type A personalities have very high levels of *commitment* in their lives. Commitment means they can persist on tasks that have consequences for them, and they are highly resistant to anything or anyone standing in their way. Third, they see *challenge* in their everyday pursuits, and challenge has an energizing impact on them. Challenge implies conflict and the need to win contests that matter. This leads us to a brief summary and conclusion about Type A behavior, which, you will recall, is associated with a very high risk of cardiac problems.

Health is endangered to the extent that control, commitment, and challenge are core aspects of an individual's personality.

Earlier on I concluded that control, commitment, and challenge were core features of hardiness and helped prevent health problems. Now I've concluded these same characteristics are associated with a major source of human anguish—coronary heart disease. In short, we face a potentially deadly paradox: control, commitment, and challenge in our lives both enhance and endanger our health. What's going on here?

Probing a Paradox: The Subtleties of Personality and Health

There are several issues we need to deal with before resolving this apparent paradox. Let's start with the issue of surface versus deep features of Type A personality. I mentioned earlier that many of the characteristics that have been shown to differentiate Type A personality from others are primarily surface traits, with the implication being that there may be something deeper at play in their personalities that is the "behavioral pathogen" for cardiac risk. Is this true?

There is convincing evidence that *hostility* is the core pathological feature underlying the different aspects of Type A behavior.[5] Hostility, in short, can kill you. Consider everyday events in which an angry or hostile response is a real possibility. The traffic is ridiculously slow today—why can't those idiots realize that a yellow light DOES NOT mean slow down!? The elevator is unbearably slow this morning—it must be those jerks on the twenty-third floor who keep opening the damned door for each other, or this great thundering twit in the long line at the grocery story cash register is unable to decide on her ridiculous Lotto ticket number, and so on, *ad nauseum*. Literally.

Recognizing that it is the deeper trait of hostility rather than the surface traits of hurry sickness or workaholism that poses health risks has important practical implications. Consider that you are the spouse of someone you suspect is Type A and assume he is a man (yes, women can also be Type As, but are less likely to be). You have decided on a Caribbean vacation, the first you have taken and something you have saved and planned for over many years. You arrive at the resort, head down to the beach, stretch out on the sand, and succumb to the delights of sun, surf, and strange drinks with exotic (and erotic) names and little umbrellas. Your Type A husband has been a sport. He has arrived with several file folders from work as well as his laptop. But he has agreed to leave them in the hotel room and only deal with them if an urgent matter arises. After three minutes he suddenly remembers a magazine he had meant to bring down to the beach and bounds off. But you know what's up. When you get back to the room an hour later there he is, furtively answering his e-mail because without him, so he says, the company will topple.

What are you tempted to do? I suspect that for many spouses this might be seen as the propitious time for a deeply serious talk. You are tempted to confront him and say that for once he should bloody well relax. This is a vacation, and he's ruining it for you. And you tell him he's ruining it for himself and his health, and he'd better get back to the beach and relax—or else! Is this precisely what is needed to safeguard the health of a loved one?

No, it isn't. This is precisely the wrong thing to do. Remember that time urgency, a hectic pace of life, and the desire for control are not the pathogens; rather, it is the hostility that underlies them. He may be hostile underneath, or he may not. And if not, then pushing him to relax when he is obviously deeply committed to his work is likely to cause hostile feelings, which are precisely what we wish to avoid. And if he is already a hostile Type A, the situation is even worse. There are other ways of handling this kind of spousal challenge, and I'll let you in on one before we finish this chapter. But first, how can we resolve the paradox about hardiness and Type A personalities?

Resolving the Paradox: A Sense of Perspective

Let's consider the three aspects of hardiness and Type A personality that give rise to the paradoxical conclusion that control, commitment, and challenge both enhance and endanger our health.

Control, as we saw in the last chapter, is a complex concept. When considered in the context of hardiness, a sense of control refers to people asserting influence over important events that matter to them. But when considered in the context of Type A personality, control has a more manipulative and maladaptive aspect to it. Recall our earlier discussion in Chapter 1 about hostility involving the extortion of validation for a construct that one believes may already be invalidated? I think that such an apprehension occurs in the case of Type A behavior. Indeed, an early assumption about Type A behavior was that individuals high in Type A tendencies have a low level of self-esteem, and their need to seek out and defend their sense of control is an attempt to deal with challenges to their self-esteem. In contrast with the indiscriminate control Type A personalities exercise, hardy people employ a sense of control that is more flexible and calibrated. In the language of Chapter 5, hardy individuals are aware of their buttons and know when to push them and when to adopt other strategies.

The issue of commitment is rather more complex. I think that Type As can adopt one of two strategies that I call *hyper-commitment* and *myopic commitment*. Hyper-commitment is a tendency to invest strongly in every task or project that presents itself. Such investments are not screened in terms of whether they are valued or valuable courses of action. This strategy also poses problems of project overload in the Type A's life, and as more and more commitments build, the costs of dealing effectively with any one of them diminishes sharply. Myopic commitment, in contrast, involves an *idée fixe*, a focusing of all of one's energies and passions to the pursuit of one overriding goal to the exclusion of everything else. In contrast to both of these Type A approaches, hardy people's sense of commitment is more discriminating, and they are able to focus their energy and attention as needed, depending on the events and concerns they are facing.

Type A and hardy people also approach challenge differently. For Type As there can be a remarkable range of events and tasks that elicit a competitive and challenging response. A colleague of mine tells the story of his five-year-old daughter being driven to school one morning by her mom and suddenly asking, "Where have all the idiots gone?" To which her mom softly replied, "They only come out when your dad is driving, dear." Lane switching and line jumping brought about by conspiracies of moronic drivers illustrates how Type As' eternal vigilance and readiness to do battle extracts a high toll on health. Hardy people, in contrast, approach the challenges in their lives without the grim earnestness of the Type A. I suspect they have the capacity to think of challenges as games, not in any trivial sense but rather in a fully engaged and enthusiastic, even playful fashion. And it is this nonhostile way of engaging with challenging events that serves as a health protective factor.

When I used to coach my kids' soccer teams the antics of the parents on the sidelines intrigued me just as much as those of the eleven-year-old kids on the field. I remember one game in particular when we were down by three goals, with time running out in the second half. Ricky's dad, Gus, was a truly obnoxious man. His typical expression during games was a sour-faced glower even when the team was doing very well.

But when the team was losing? He took such situations as a personal insult and let his son know it by screaming at him, "For Chrissake, Ricky, HUSTLE! This isn't a goddammed game." When I pointed out to Gus that this *was* a game (goddammed or not) he was not amused. And the fact that our exchange was carried out in front of all the other parents simply ratcheted up his hostility even further. He was apoplectic.

In contrast to Gus, most of the other parents were engaged and involved; in fact, they were super-keen, but when things would get tense during a game they responded not with Type A yelling but with hearty and, I suspect, hardy cheers for their kids. What I found particularly interesting was that they were often smiling during the games. Not only that, but they displayed what we call Duchenne smiles, the genuine ones that animate the full face, not just the lips, in contrast with the forced, non-Duchenne smile of those who are faking it.

What can we conclude, then, about the apparent paradox of control, commitment, and challenge? I think the major difference between Type As and hardy individuals is one of *perspective*. Because hostility underlies much of what Type As experience, each of these otherwise adaptive orientations is carried to an extreme that can push the autonomic nervous system into overdrive and increase the likelihood of stress-related health decline. It is also true that these differences in perspective may be linked to fairly stable features of personality; recall that in Chapter 2 I mentioned that low scores on the Big Five trait of agreeableness were also strongly associated with increased risk of coronary disease. I also mentioned that there are some steps that can be taken to avoid some of the triggers that activate Type A behavior. Although this is definitely not intended to be a conventional self-help book, it might be helpful to give an example of a strategy that works well for most people and worked for me—well, almost worked for me.

SO WHAT THE HELL CAN WE DO ABOUT HOSTILITY?

I had been carrying out research on personality and health for a few years when I had the opportunity to attend grand rounds at a large psychiatric hospital where the demonstration for that month was about "thought stopping" for reducing hostile behavior. There was a group of fifteen of us in all manner of dress, professional background, and levels of hostility. The demonstrator asked us to close our eyes and imagine, for about three minutes, a scene that made us frustrated and at least somewhat hostile. I conjured up a vision of the same idiot drivers that my Type A friend had spotted so readily and who seemed also to turn up occasionally on my drive to the university. I had been deeply concentrating on this image and building up a nice wad of anger when the demonstrator's voice screamed through the microphone: *STOP!!!* We all jumped and stopped simultaneously. Our instructor then asked us how many of us were still thinking about the event that made us hostile. None of us were. In my case the *STOP* interruption had been sufficiently intrusive to totally divert me away from thoughts of road rage and digital gestures.

The next step in our demonstration was to show how we could incorporate this "cue word"—STOP—whenever we wished to be diverted from a course of thinking that was creating hostility, anxiety, or any other undesirable emotion. It would have been helpful to have had the demonstrator permanently attached to us to yell appropriately when he thought we were experiencing unpleasant emotions—sort of a Global Psychiatric Screamer, or GPS—but that seemed impracticable. So we were trained to interrupt ourselves with the word *Stop* when we knew we were in a situation in which hostility or anxiety was about to accelerate. At first we did it out loud, but we quickly learned to internalize the *Stop* so that only we were aware of the interruption signal. Over the next few weeks I found several occasions during which I was able to interrupt an unwanted state of mind by simply intoning "Stop" sharply but silently. I also found it was more effective if I quickly blinked my eyes once while saying it to myself.

It turned out that within a week of sitting through the grand rounds demonstration I had occasion to use the thought-stopping technique. My dean had asked me to tackle a problem of accountability in evaluating the university's courses of instruction. The logic was that the government was going to impose its own process of figuring out how well universities carried out their tasks, so preempting them by developing our own standards would be highly desirable. This meant I would need to visit each department in the university and brief them on our system of self-accountability. I knew that universities were notoriously resistant to change and that changing fundamental aspects of the academy was like moving a cemetery. So when asked to do this, I said that I would have preferred to have open rectal surgery without an anesthetic, to which the dean responded, "That can *also* be arranged, Professor Little."

The night before my first briefing I received an e-mail that confirmed my expectation of trouble ahead. The e-mail was in response to a report I had written outlining some procedures for assessing accountability, and a very senior professor in the economics department expressed his concern directly: "I am appalled by your report. I will see you at the meeting tomorrow. Be prepared to fight." When I went to bed a few minutes later I guess this unexpected and not overly collegial e-mail had me rather stressed out. My wife asked me if I was tense, perhaps because I was *standing* up in bed. I told her what was up, and she suggested that I practice what I had recently learned about thought-stopping techniques. We agreed that I needed a keyword that would allow me to stave off the stress likely to emerge as an out-of-control, Type A macroeconomist savaged me. She suggested I just make a subvocal quacking sound to evoke the image of water rolling off a duck's back. I thought it was a brilliant suggestion and drove myself to the university the next morning with a new honking sound in my behavioral repertoire.

When I got up to defend my position at the meeting my combatant rose simultaneously, and we stared at each other. I silently quacked, all sense of anxiety dissipated, and I calmly and solicitously said that Professor X had something very important to say. The economist, somewhat startled, quietly made his point and sat down. Throughout the rest of the meeting I managed to defend a highly controversial policy with something very close to aplomb, thanks to my subvocal quacking.

But toward the end of the session a psychologist and a political scientist were debating the issues, and I intervened to clarify some obscurities in the political scientist's remarks. He wheeled around, glared at me, and shouted, "Professor Little, when I want my position clarified I'll ASK TO HAVE IT CLARIFIED!" It took me aback, but my newly acquired thought-stopping skill kicked in. To my considerable embarrassment, however, what I thought was a subvocal sound clearly passed the threshold into a highly vocal one. I let out a very loud QUACK. The political scientist look puzzled and asked, "Brian, did you just quack?" I then performed an extrication procedure that was pretty pathetic. I pretended that I had just coughed and repeated it four times with a distinctly quacking sound accompanying it. This was not my finest moment.

The moral of this story is that one can handle hostility, anxiety, and other responses to stress in the short term through techniques cognitive behavior therapists use, like thought stopping or strategic relaxation. But sometimes attempts to suppress a response can backfire. Recall our earlier account of Dan Wegner's work on ironic processes and the process of not thinking about a white bear (or green cat)? The suppressed thought becomes *more* rather than less likely to appear in consciousness.[6] Once again, try it yourself, take three minutes right now and try not to think of a duck. Seriously.

A SENSE OF COHERENCE: PERSONALITY, HEALTH, AND CONTEXT

In the last two chapters we have been concerned with the relation between a sense of control, or agency, in our lives and diverse forms of well-being, including academic and occupational success as well as physical health. We concluded that a sense of control in our lives provided clear benefits, but it also needed to be based on an accurate reading of actual control. We saw how hardiness also has a salutary effect on individuals who have been exposed to stress, but that some of its elements, if carried to an extreme, as with Type A personalities, may enhance rather than diminish risks to health. There is one final concern to consider that provides not only a valuable perspective on how personal dispositions shape the course of life and well-being but also how our environment plays a vital role in the process.

This integrative theory was proposed by Aaron Antonovsky, a medical sociologist who distinguished between the pathogenic view of illness, which traces the origin of disease, and what he called the *salutogenic* process, which examines the source and development of health. Central to salutogenesis was a person's *sense of coherence* (SOC), defined as "the extent to which one has a pervasive, enduring though dynamic, feeling of confidence that one's environment is predictable and that things will work out as well as can reasonably be expected."[7] SOC is based on three elements. Comprehensibility is the extent to which one's daily life is seen as making logical sense and is ordered and predictable. Manageability is the extent to which one feels able to cope with environmental demands. Meaningfulness is how much one is able to invest in daily projects and pursuits as endeavors worthy of commitment. People high on SOC have more of what Antonovsky called "generalized resistant resources," which enable them to remain mentally and physically healthy after being exposed to challenges.

I particularly like this phrase in the definition of SOC: "as well as can reasonably be expected." It strikes the same tone that we invoked in talking about checking our buttons before assuming total control and in discussing how hardy individuals, in contrast with those with Type A personalities, approach challenges in their lives. But the notion of a sense of coherence has an important additional message about health and well-being: it explicitly shines a spotlight on the nature of our environments, our communities, as critical components of coherence. One of the best examples of this was a study carried out in three different communities in Israel: one was a traditional, agrarian community; one was a modern city; and the third was transitional between the other two. It was predicted that the sense of coherence and health status would differ between the three types of community. Take the notion of manageability as an example. In the agrarian villages a sense of the predictability of daily life is passed down through traditional practice. Here it is not so much the feeling that one is in control but that "things are *under* control" that contributes to a sense of coherence. In large cities, in contrast, a sense of coherence is more likely to emerge when one has personal control over everyday pursuits. Those who live in transitional settings had neither the stability provided by tradition nor the agency afforded by modernity, so residents of such settings would be expected to have a lower sense of coherence within their lives. The empirical evidence supported this prediction. When measured on a measure of SOC, people in both the traditional and urban settings scored relatively high (at about the same level), and those in the transitional setting scored significantly lower. And consistent with the theoretically anticipated health effects of SOC, it was in the transitional settings that health problems were more prevalent.

We might speculate that drops in a sense of coherence and ensuing health declines may occur in other transitional periods during our lives. Leaving home, changing jobs, falling in and out of love, having a child, retiring—all average and expectable events in many lives—may cause temporary shifts in our sense of coherence. We also know that it is during these transitional periods that differences in personality are most likely to be manifested.[8] During a transitional period extraverts are likely to be particularly extraverted, conscientious people are even more orderly and organized, and disagreeable people are especially unpleasant. Research on the sense of coherence also raises some fascinating issues for designing settings and communities that are comprehensible, manageable, and meaningful. This means knowing how personalities and places interact so lives can be enhanced . . . as well as can reasonably be expected. Or perhaps even more.

chapter seven

Personality and Creativity: The Myth of the Solo Hero

> There is nothing more difficult to pull off, more chancy
> to succeed in, or more dangerous to manage,
> than the introduction of a new order of things.
>
> NICCOLÒ MACHIAVELLI, *The Prince*, 1532

> When you make music or write or create, it's really your job
> to have mind-blowing, irresponsible, condomless sex with
> whatever idea it is you're writing about at the time.
>
> LADY GAGA[1]

T HINK OF THE MOST CREATIVE PEOPLE YOU HAVE ENCOUNTERED in your life. Some you may know only indirectly, through experiencing their creative works. You have been absorbed in their novels, addicted to their video games, danced alone to their music, or pleasantly befuddled by their performance art. Others you know more directly: your physician who devised a new regimen for your child that actually worked, your plumber who fixed a deeply neurotic sump pump when no one else could, your second spouse who helped you pick up the pieces and to see them in a revealing new light. Do such individuals share anything in common? What differentiates them from those who adopt more conventional ways of solving life's problems and challenges? Are *you* creative, and will you still wish to be after you read this chapter?

You might be interested in doing a brief assessment that will be helpful in making what follows more personally relevant to you.

Check off any of the following adjectives that you believe accurately describe you:

☐ affected	☑ intelligent •
☑ capable •	☑ interests narrow -
☑ cautious -	☐ interests wide
☐ clever	☐ inventive
☐ commonplace	☑ mannerly -
☑ confident •	☐ original
☑ conservative-	☑ reflective •
☑ conventional -	☑ resourceful •
☐ dissatisfied	☑ self-confident •
☐ egotistical	☐ sexy
☑ honest -	☑ sincere -
☐ humorous	☐ snobbish
☐ individualistic	☐ submissive
☐ informal	☐ suspicious
☑ insightful •	☐ unconventional

This scale was developed by Harrison Gough of the University of California, Berkeley, and has been used extensively for research purposes as a brief, valid measure of creative personality.[2] To obtain your score, add up the total number of checks you gave for the items *capable, clever, confident, egotistical, humorous, individualistic, informal, insightful, intelligent, interests wide, inventive, original, reflective, resourceful, self-confident, sexy, snobbish,* and *unconventional*. Now subtract from this total your number of checks on the items *affected, cautious, conservative, conventional, dissatisfied, honest, interests narrow, mannerly, sincere, submissive,* and *suspicious*. The theoretical range of scores is therefore from −12 to +18. Scores of 10 or more are similar to those of highly creative individuals.

THE IPAR CREATIVITY STUDIES

The Institute for Personality Assessment and Research (IPAR) at the University of California, Berkeley, was originally located in a renovated fraternity house on a wooded street at the top of the UC campus near Grizzly Peak Boulevard. I have a vivid image of writing my PhD comps in the summer of 1967 in a small second-floor room at IPAR, with the smell of eucalyptus wafting through the open window and both coffee and creativity percolating across the hall. Donald MacKinnon, the director of IPAR, had an office a few steps away, and I recall the first time I met with him. I was describing my research ideas, and he was considering being my adviser. He was not an overly expressive man, and he read over my research prospectus without giving any hint of what he thought. He then looked up, paused, and said, "Little, are you a doer or a thinker?" My immediate response was "I'm a doer, I think," and the answer seemed to satisfy him, although my emphasis on the word *think* conveyed my conviction that thinkers can also be doers. It wasn't until later that I realized that these two attributes—thinking, particularly innovative thinking, and doing, the ability to convert those ideas into action, were the key criteria IPAR used to identify and study outstandingly creative individuals who had transformed their fields. The extensive set of studies done at IPAR in the early 1960s transformed our understanding of personality and creativity.[3]

Look What I Did! Norms, Narcissism, and Creative Displays

How do we know that someone is creative? The most prevalent view among researchers in this area is that creative individuals are those who have been responsible for generating products that are both novel and useful. These products may be ideas, objects, or processes. Mere novelty is insufficient for the designation of creativity, or else all manner of strange but useless novelties would be called creative. Nor is mere utility sufficient to be called creative; there has to be both. And the evaluation of both innovation and utility is a normative judgment. It is based on comparisons with what have been the conventional norms of a particular domain, be it French cooking, organic chemistry, gangsta rap, or architectural design.

But at the outset this definition of creativity raises a tricky issue, particularly when we are judging creative products and creative people on short periods of firsthand experience with them, such as in a typical job interview or in brief presentations to clients or potential customers. Here's the problem: What if the person you are interviewing or evaluating is a narcissist? Recent research shows that narcissistic individuals evaluate their own products and projects as being highly creative.[4] Indeed, narcissists invest greatly in standing out from others, and one way of doing this is through investing in projects that display themselves as distinctive. These resemble mating displays in certain avian species. I have seen more than a few narcissistic academics advertising their prowess at conferences with as much subtlety as an overly aroused peacock displaying its tail feathers: "Look at my vita! Look at my vita!" However, what the research shows is that narcissists, despite believing themselves to be creative, actually are not creative, at least according to objective tests. But this isn't just *self*-delusion; they are also skilled at convincing *others* that they are creative. When asked to pitch ideas for Hollywood movie scripts, for example, narcissists show greater enthusiasm and charisma in their pitches. This capacity to convince themselves and others about their innovative worth might lead them to be erroneously regarded as truly creative. Particularly in domains in which there are few objective criteria for novelty or value, it is better to look at long-term evaluation of a succession of innovative and influential hits rather than a single exposure crafted by a narcissistic pitcher.

IPAR did just this. They drew on the judgments of individuals who were experts in particular domains to help select people who had a long history of creative achievements in their particular fields and who had created new norms of excellence within those fields. They studied diverse groups, including novelists, scientists, managers, military officers, mathematicians, and graduate students. But it was their assessment of creative architects that became the best-known and most influential of these studies.

The first task—and not an easy one—was to determine who qualified as being the most creative architects in North America, not simply the most productive architects. The selection criteria were simple but stringent. Those who would comprise the creative group to be invited to Berkeley were to have accomplished three things: they had to have devised novel, innovative forms of architecture; these novel forms must have been actually implemented—they needed to be both thinkers and doers; and these creative products must have contributed to a new standard of creative excellence in the field.

But who would best evaluate architects on these criteria? Creative achievement is evaluated as such by the consensus of those who are expert in a particular domain of practice. I have a pretty good idea about whether my students' ideas in personality psychology are creative, but I am way out of my depth if asked to judge the creativity of hairdressers, professional yodelers, or morticians. Each field has its guild of knowledgeable experts who evaluate the originality of products in their domain. Accordingly, the IPAR staff decided to have experts in the architectural field make those initial nominations of individuals who had transformed their fields. Each was asked to nominate a set of highly creative architects using the three criteria of novelty, implementation, and standard setting. It was recognized at the outset that this might backfire. It was possible that each editor would nominate a different group of creative architects. Fortunately, this was not the case. There was a sufficient degree of agreement so that a group of forty architects was identified whom subsequent groups of architectural judges confirmed had creatively transformed their field.

If IPAR had simply studied the characteristics of this creative group, the results might have simply reflected the fact that all of them were highly functioning architects, most living in major metropolitan areas and well engaged with their professional communities. What was needed was a control or comparison group of architects who shared these characteristics but were not themselves creative. A particularly nice aspect of the IPAR research design was that they also studied architects who worked in the same firms and cities as the creative group but had not been rated as creative. They formed an excellent "control" group with whom the creative architects could be compared.

IPAR invited the creative groups to visit the Institute for three days of assessment in groups of ten. Although they stayed overnight at a nearby hotel in the Berkeley Hills, they spent each very full day being assessed by IPAR psychologists who interviewed them, administered a diversity of tests, and provided ratings of their performance on various tasks. They were also observed and evaluated on how they behaved in the social situations at lunch and in between assessment tasks. Some of the tasks they were asked to complete were deliberately made to be stressful so their ability to perform under pressure could be assessed. It was a pretty intense weekend, and not all participants regarded it as a delightful experience. Indeed, one poet, Kenneth Rexroth, who participated in one of the other IPAR creativity studies on writers, had some decidedly unkind things to say about his experience. He wrote up his experiences in a grumpy, funny, and decidedly mocking article called "My Head Gets Tooken Apart."[5] Studying creativity by assessing those who are creative can be inordinately challenging; they don't want us to mess with their magic. Others, as we might expect, rose to the occasion at IPAR, threw themselves into the project, and enjoyed the experience.

The overarching research question was this: How did the highly creative architects differ from their highly capable but less creative peers on measures of ability, background and early experience, personality, and social functioning? Because the results of the architects' study were largely replicated with other groups, in what follows I will talk about creative people in general but, where appropriate, will refer specifically to the architect study.

One of the hazards of giving talks or writing books about psychological research is that occasionally listeners or readers will respond to the presentation of findings by saying something like, "Sure, sounds about right" or "Of course" or "Everyone knows that." This can be disconcerting not only because it could be true but also because it is possible they would not have predicted the results beforehand. So, when lecturing, I often ask my audience to predict *beforehand* what the results will be of the studies I am about to relate. It is always more compelling to bet on a horse before the race is run rather than after it.

So in the spirit of sharpening your anticipation, let me ask you to make some predictions regarding the personality and life experiences of highly creative individuals relative to more conventional ones. These questions only cover a few of the issues we discuss in this chapter, but they touch on a number of the key findings. Are highly creative individuals more intelligent or less intelligent than conventional individuals? Are they more likely to have had emotionally intense relations with their parents or less likely? Are they likely to have excelled in school or to have done relatively poorly in school? Are they more likely to have the same interests as bankers or as lawyers? Are they more likely to be extraverted or introverted? Do they prefer chaotic complexity or elegant simplicity? Are they more likely to be vulnerable to mental illness or more likely to be mentally stable? And as you think about these questions you may also want to think about how you would score on these different traits, preferences, and orientations.

Intelligence and Creativity: Simply Brighter?

Are highly creative individuals simply brighter and more intelligent than their conventional peers? The IPAR study showed that this was not the case. There were no significant differences in IQ between the two groups. However, remember that all participants in the IPAR studies, regardless of whether they were creative, were highly educated professionals. Beyond an IQ level of about 120, which is typical of highly functioning professionals, higher levels of IQ are unrelated to creativity, so someone with an IQ of 145 has about an equal chance of being either creative or conventional. It should be noted, however, that conventional IQ tests discriminate most reliably among individuals at the average IQ level of 100 and don't do as reliable a job in differentiating between people at the extremes. Had they used a conventional IQ test, then, the finding of no difference between conventional and creative architects might have been attributable to this lack of sensitivity in measuring IQ. However, the IPAR researchers used a special assessment tool called the Terman Concept Mastery Test that was specially designed to discriminate most reliably at an IQ level of 120. It was on this finely calibrated measure of intelligence that creative and conventional groups were found to score the same. Those at the very highest level of creativity in their fields, in short, are bright, but they are no brighter than their less creative peers.

In high school were those who would become creative innovators better students? Not in the conventional sense of getting straight As. In fact, they typically graduated as B students. A frequently found pattern was getting very high grades in courses they identified with and very mediocre grades, if that, in courses with which they found no sense of connection.

Early Experiences: Developing Degrees of Freedom

Several converging themes were found in the highly creative individuals regarding their early experiences and education. In early childhood their families accorded them a great deal of respect and allowed them to explore on their own and develop a strong sense of personal autonomy. There was also a lack of extreme emotional closeness with parents. There was little evidence of intensely negative experiences; for example there was, relative to the times in which they lived, very little physical punishment for transgressions. Nor, on the positive side, was there evidence of extremely intense bonds of the sort that can smother independence. On balance, for those who would grow up to be highly creative, relationships with parents were relatively easy and, in later life, pleasant and friendly rather than intensely intimate.

A similar pattern was found with respect to the role of religion in childhood. There were no denominational differences between creative and more conventional architects, but in the case of the creative architects' backgrounds there was more of a focus on helping the child develop an internal code of values rather than strict adherence to doctrinal aspects of religion.

The creative group also experienced a significantly larger number of moves while they were growing up. This is likely to have increased their degree of adaptive flexibility in contrast to the experience of staying in the same location for a long period. In personal construct theory terms, they were likely to have developed more complexity in their personal construct systems. But the frequency of moves may have also contributed to a sense of isolation from other individuals. Instead of the stabilizing support of long-term friendships, they came to rely more on their own resources.

The converging pattern that we see in highly creative individuals' early influences and experiences is the encouragement of individuality, personal autonomy, and far greater degrees of freedom from the kinds of emotional and intellectual constraints that would be found in the lives of more conventional people.

Interests and Orientation

One of the tests administered in the IPAR study, the Strong Vocational Interest Blank (SVIB), measured the similarity between participants' interests to those of a large array of occupational groups for whom there were extensive norms.[6] The highly creative groups' interests were more like those of psychologists, author journalists, lawyers, architects, artists, and musicians. Their interests were very *unlike* those of purchasing agents, office workers, bankers, farmers, carpenters, veterinarians, police officers, and morticians.

This pattern of interests suggests that highly creative individuals are not interested in facts for their own sake but instead in their meaning, significance, and implications. They are inclined to see emergent forests rather than isolated trees and have an affinity for skilled communication about such ideas. They are rather averse to conventional and highly regulated activity. They can get exasperated with details. Their interests suggest that they are cognitively flexible, verbally sophisticated, and intellectually curious. They are disinclined to police their own impulses and ideas and, perhaps, those of others as well.

Reflecting the demography and norms of the day, the group of architects studied at IPAR were all male.[7] One of the most intriguing findings with the creative group concerned whether their interests tilted in the direction of more masculine pursuits or those that were coded as more feminine. The coding was taken from the SVIB manual, which showed which interest items differentiated between the women and the men in the norm group. The evidence is very strong that the creative architects have considerably more femininity of interests than masculine. This result also held with the other groups studied at IPAR.

This result deserves a closer look. If we examine the specific items that empirically differentiate women's and men's interests, many of them concern activities like attending concerts or going to art exhibits, which might be better regarded as cultural interests rather than feminine in a restricted sense. It is also important to note that the SVIB femininity scale is scored such that the items get coded in terms of a single dimension of masculinity at one end to femininity at the other. It was impossible, therefore, to gain high scores on both masculine and feminine interests on the scales used at IPAR. In subsequent years techniques were developed that assessed psychological masculinity and femininity as independent orientations. There is good reason to believe that the highly creative architects would have scored high on both orientations.[8]

Patterns of Preference

A similar pattern is seen in the creative architects' responses to the Myers-Briggs Type Indicator, in which contrasting preferences for apprehending the world are assessed.[9] The first contrast distinguished between introverted and extraverted orientations to the world. Highly creative individuals assessed at IPAR are consistently more likely to score as introverts; indeed, two-thirds of creative groups studied at IPAR are introverts, considerably more than the general population.

The second contrast concerns the way in which one construes information that arises out of external and internal sources. Two attitudes can be distinguished. One can perceive such events by being aware of them and open to their meaning and significance, or one can judge them by drawing a conclusion about them. One potential problem of the judging orientation is that one may *pre*judge events and draw conclusions that, although they may provide order in one's life, subvert the possibility of learning something new. In this contrast between perceiving and judging, creative groups consistently tilt in the direction of the perceiving attitudes. Although this orientation leads to greater attentiveness to one's internal and external sources of stimulation and a more engaged and open attitude toward them, it might also lead to a lack of order and structure in one's experiential life. Creative lives can be chaotic.

A third contrast is between two different types of perception: sensing, which involves attending to the immediately sensed reality of events and objects, and intuiting, which involves a perception of the meaning and possibilities inherent in what is perceived. There is a strong preference in the general population for a sensing orientation. People with a preference for sensing might be characterized as "get real" people, and they can be baffled and impatient with the intuitive people who go beyond immediate facts to imagine potentials and possibilities.

Not surprisingly the highly creative groups studied at IPAR displayed an exceptionally strong preference for intuition. Whereas an estimated 25 percent of the general population is intuitive, among the highly creative groups assessed at Berkeley, 90 percent of the creative writers, 92 percent of the mathematicians, and 93 percent of the research scientists were intuitive. Strikingly, the MBTI measured 100 percent of the creative architects as intuitive.

A fourth contrast revealed on the MBTI concerns differences in whether individuals judge events and objects through thinking or feeling. A thinking orientation evaluates and judges on the basis of logic and rational analysis, whereas a feeling orientation is based on appraising one's emotional reaction. In this case the specific field of the creative individual was an important factor: the thinking-feeling distinction applies differently to those who are creative in scientific versus artistic and literary fields. Creative scientists score higher on the thinking orientation, whereas creative writers score higher on the feeling orientation. Interestingly, the creative architects split fifty-fifty in terms of their preference for the thinking or feeling function.

Engaging Complexity, Elegant Simplicity

When I used to lecture on creativity with my small seminar classes, just before the half-time break in one of the lectures I would sing out, very loudly, "DUM DIDDILY UM DUM," but without the traditional "DUM DUM" at the end. It was fun to watch the students' responses. I did this not to be perverse but to give them a foreshadowing of an issue I would be lecturing on in the second half of the lecture. That topic was the difference between highly creative and less creative individuals in terms of their preference for aesthetic experiences that were complex, asymmetrical, and tension inducing versus simple, symmetrical, and tension reducing. The studies at IPAR examined precisely such preferences, but in the domain of visual aesthetics. Would highly creative people prefer the visual equivalent of an open-ended, asymmetrical DUM DUM–less riff?

Imagine that you have been given an eight-by-ten-inch board and a large selection of variously colored one-inch squares and asked to create a pleasant, completely filled-in mosaic within thirty minutes. What kind of mosaic would you construct? When the creative groups were asked to do this task they showed a clear preference for constructing complex, asymmetrical patterns in contrast with the simpler, balanced, and symmetrical mosaics of the more conventional individuals.

Similar results were found with a test designed to assess preferences for various kinds of pictures. Participants were asked to sort 102 postcard-size reproductions of European paintings into four preference categories. In all of the IPAR studies the creative groups expressed very strong preference for more complex, asymmetrical, unbalanced paintings in contrast with their more conventional colleagues. But something is missing here. Although highly creative persons' preference for complexity will be apparent at the beginning of a creative project, during its unfolding there will be a strong motivational drive that then culminates in a creative resolution of the complexity. This might give the appearance of simplicity, but it needs to be appreciated as a process extended over time that grapples with complex and demanding questions and then resolves them in an unconventional but elegant fashion with a clear and concluding DUM DUM!

Creative Personalities: Odd or Audacious?

When asked to describe themselves there were striking differences between creative and conventional architects' personalities that were also found in other groups of people assessed at IPAR. Highly creative individuals described themselves as inventive, determined, independent, individualistic, enthusiastic, and industrious. More conventional individuals described themselves as responsible, sincere, reliable, dependable, clear thinking, tolerant, understanding.

On the California Personality Inventory (CPI), a test that provides a detailed assessment of individual differences in personality, a very similar picture emerged, and it is useful to look at the results of the comparison between the creative and less creative architects in some detail. Here is a detailed description of the creative group as written by Harrison Gough, the psychologist who developed the CPI:

> The highly creative person is dominant; possessed of those qualities and attributes which underlie and lead to the achievement of social status; poised, spontaneous, and self-confident in personal and social interaction; though not of an especially sociable or participative temperament.

Taken together with the evidence that creative individuals tend to be introverted, the CPI results paint an intriguing picture of their typical stance toward other individuals. Creative individuals might be regarded as asocial: they are neither drawn to interactions with others nor strongly antagonistic to them. Rather, their passions are concentrated on the domains in which they pursue their creative projects. This might well give rise to the impression that they are standoffish and rather arrogant. But when the situation calls for it the highly creative individual has the social poise and social skill to be charming and even endearing. Yet sustained sociability is not the mark of most creative individuals, and this can cause friction with colleagues and friends who resonate with the charisma but want it to last more than a few fleeting moments.[10]

> With respect to intellectual style, the highly creative person is:
>
> Intelligent, outspoken, sharp-witted, demanding, aggressive, and self-centered; persuasive and verbally fluent, self-confident and self-assured; and relatively uninhibited in expressing his worries and complaints.

In short, highly creative individuals can be extremely demanding on those with whom they work. They can terrify the timid, sometimes without realizing it, through sheer force of personality.

The potential difficulties of working with creative individuals also are seen in their discomfort with the conventional and their willingness to be audacious and even somewhat odd.

> He is relatively free from conventions and inhibitions, not preoccupied with the impression which he makes on others and thus perhaps capable of great independence and autonomy and relatively ready to recognize and admit self-views that are unusual and unconventional. He is strongly motivated to achieve in situations in which independence in thought and action are called for. But, unlike his less creative colleagues, he is less inclined to strive for achievement in settings where conforming behaviour is expected or required.

Given their personalities, it is easy to imagine the difficulties organizations would have with a highly creative colleague. Because they are unconstrained in their individuality and have no particular desire to create a good impression, they run the risk of subverting business as usual, at least in the conventional sense. In situations in which sustained tact, diplomacy, and give-and-take are required, the highly creative person can wreak havoc. Although they can be charming and charismatic, they can also be bloody minded and unable to rein themselves in, despite their colleagues' panicky entreaties. Brilliant administrators will often arrange to have the creative types strategically diverted from meetings in which the agenda calls for conventional problem solving or the careful and judicious weighing of options. There is little doubt that, love them or hate them, creative individuals are audacious. But are they also odd? Is there any truth to the age-old view that creativity and madness are closely aligned?

STRANGE CREATURES? CREATIVITY, ECCENTRICITY, AND PSYCHOPATHOLOGY

It is important to differentiate between individuals who are creative, eccentric, and mentally ill. They share some features but differ importantly in others. We'll take a slight diversion to look at eccentricity and then deal with whether the creative groups at IPAR were, in some sense, manifesting mental illness.

Eccentrics: Happy Obliviousness

David Weeks and Jamie James have presented a compelling account of eccentrics and have given us some illuminating and entertaining pictures of their colorful behavior.[11] One of the most striking is the case of Joshua Abraham Norton, who in the mid-nineteenth century was, he claimed, the Emperor of the United States and Protector of Mexico. Norton was accorded remarkable privileges in San Francisco (of course), where he strode the city in full military blue uniform, replete with plumed top hat and sword. He delivered edicts galore, believed he had dissolved both the Republican and Democratic parties, and issued his own currency, which was recognized and honored in shops and establishments around the Bay Area. Although most of his edicts were potty, some were prescient. He agitated for the completion of a bridge from Oakland to San Francisco and a tunnel under San Francisco Bay, both of which came to fruition long after his death in 1880.

San Francisco is an undeniably liberal, if not totally loopy, city—an ideal milieu in which Emperor Norton could fully engage his fancies and be treated with tolerance and fondness. In other cities (you know where they are) he would not have fared so well. Today he would likely cause major delays at the security lines of airports, and not just because of his sword. So the nature of the setting in which eccentrics are nourished and flourish is important. It is worth noting that Joshua Abraham Norton was born in London, England, a place famously tolerant of eccentrics of all kinds.

Here is my own experience with an English eccentric, one who might have given Emperor Norton a run for his very own money. I spend several months each year in Cambridge, England, where there is no shortage of eccentrics. One elderly woman, whom I see frequently, rides a high handlebar bicycle at a terrifying speed through the cobblestone streets. She dresses in a Muammar Gaddafi–style dress military uniform, a brightly plumed hat, and red sneakers. I call her Maude. A very loud whistle seems to be lodged permanently in Maude's mouth. Whenever she sees something that strikes her as annoying she blows her whistle very loudly and repeatedly until the miscreants get out of her way or stop looking strangely at her. She sometimes comes close to Armstrong levels of velocity when she gets aroused, and I bet her speed is not steroid enhanced. Maude doesn't just steer her bicycle; she aims it. I saw a whole busload of Chinese tourists scattered like bowling pins when they dared to cross the street as she was hurtling down it.

Maude is clearly eccentric, but without further information it is difficult to say whether she is creative or deeply disturbed. Given that it is Cambridge, it is entirely possible that she is simply a dotty professor emerita, trying to save her beloved university from the unrelenting hordes of visitors. But I have heard her blow her whistle at an ATM as well, so perhaps she is more disturbed than simply disturbing. One way of differentiating eccentrics from mentally ill individuals is that eccentrics are generally happy, sometimes exceptionally so, with their lot in life, even if most people view that life as strange and perplexing. Despite her aggravated assaults with her whistle, I suspect that Maude is rather oblivious to the impact she has on others and that she is, in her own singular way, rather content.

Norton, Maude, and other eccentrics are not necessarily mentally ill, although they can be. Nor are they necessarily creative, particularly if we regard creativity the way IPAR did, as requiring the conversion of innovative, nontraditional ideas into adaptive solutions for challenging problems. What seems to distinguish the eccentric from the certifiably creative is their obsession with their own personal projects rather than with tasks accorded significance by the larger community. What differentiates them from those who are regarded as having mental illness is their great delight in being themselves, their happy repudiation of convention, and the remarkable degrees of freedom they have in living unconstrained. Mental illness, in contrast, is not a choice. Unlike eccentricity, it is not freely chosen, and rather than liberating a people to live their lives as they wish, however aberrantly, it poses severe constraints on one's choices. And unlike eccentricity, mental illness is typically frightening, exhausting, and depleting.

Creativity and Psychopathology: The Unfiltered Mind

We have already seen that creative individuals are not particularly interested in creating a good impression and are not inhibited when it comes to expressing their feelings, including their negative feelings. In this respect they do resemble eccentrics. But the IPAR researchers were particularly interested in whether there were signs of psychopathology in these highly creative groups. At first blush it appeared that there were such signs. The best indicator here is how the creative group compared with more conventional individuals on the Minnesota Multiphasic Personality Inventory (MMPI). This inventory measures similarity of a person's responses to those obtained for patients diagnosed as having depression, hysteria, paranoia, and schizophrenia. On these and other similar scales creative individuals score considerably higher than the general population. It is not entirely misleading, then, to say that they do appear to be not only audacious but also decidedly odd. Creative people are, in important respects, very strange creatures. But is it fair to say they are more at risk for mental illness? I think the simple answer is "no, they are not." But this is a complex issue, so my answer needs qualification.

The scores on the MMPI achieved by individuals who are functioning effectively in society, like those invited to IPAR clearly were doing, need to be interpreted differently from those obtained by individuals who are experiencing problems with life or are in mental hospitals. One indicator that differentiates creative individuals from those at risk for psychopathology is their score on a measure of what is called "ego-strength." The ego-strength scale on the MMPI was originally developed to predict who would and would not benefit from psychotherapy. Those scoring high on ego-strength are intelligent, resourceful, realistic, and able to tolerate confrontation.[12] Highly creative individuals score high on ego-strength, whereas those who are more likely to suffer psychiatric disorders score very low on this dimension.

The importance of examining ego-strength as well as the psychopathological scales on the MMPI became very clear to me when I was a consultant for an organization that had just concluded a search for a senior vice president. One of the unsuccessful candidates—let's call him Dan—was well known in the community as a highly creative, even visionary leader. In conversations with the selection committee I found out that one of the reasons he had been passed over was because of his MMPI profile that, in the rather unsophisticated terminology of the committee, had shown him to be "just plain nuts." I had been given responsibility to advise on the use of psychological tests for recruitment and was extremely skeptical about using tests like the MMPI for such purposes. So with their permission I was able to access Dan's MMPI. Ego-strength is not typically reported when MMPI results are given; it is a special scale that is primarily used by researchers. When I looked at Dan's profile, sure enough, his psychopathology scale scores were elevated. But no ego-strength score had been calculated. It turned out that when I examined the fifty-two items on the ego-strength scale, Dan's score was extremely high. In short, by not taking account of the positive, coping aspects of his personality the selection committee lost the opportunity to hire a truly creative and passionate leader. Sure, Dan was odd. But he was also bright and motivated and able to transform his original and sometimes off-the-wall ideas into creative accomplishments. Odd, yes, but also audacious.

Recent research has raised the possibility that eccentrics, creative individuals, and those at risk for psychopathology are similar with respect to their relative inability to filter out extraneous information impinging upon them.[13] In order to adapt and survive, we need to selectively filter out information that has no motivational or strategic importance for us. This capacity is referred to as latent inhibition (LI), and those who are very *low* on it include creative individuals, eccentrics, and those inclined to mental illness, particularly schizophrenia. There is an upside of having low LI, however: it opens the individual to a rich array of remotely connected thoughts and images that those with more effective filters in place would have screened out. These can be a fertile ground for creative insights, heightened sensitivity, and novel ways of seeing the world. On the downside the unfiltered mind risks becoming overwhelmed and the ability to cope compromised.

So is there something about highly functioning creative individuals with low LI that differentiates them from those who succumb to psychopathology? Evidence from Jordan Peterson and his associates provides a promising clue.[14] In studies with Harvard undergraduates they showed that intelligence and good short-term memory might well be critical factors. They propose that those with higher intellectual resources can cope with the flood of information the unfiltered mind allows. This result parallels what we have already seen with respect to ego-strength in the IPAR studies. Both intelligence and ego-strength involve the ability to face complexity and information overload in an adaptive way. Without such resources, cognitive and emotional information that has no functional utility for us overwhelms us, and we risk drowning in data.

In writing about the IPAR studies MacKinnon did allow that some of those who came to Berkeley for assessment were experiencing serious psychological problems, but they were very much a minority. Highly creative individuals, in short, have vulnerabilities but generally are able to transcend them and even recruit them into the creative process. Having a stable, pleasant, happy, shiny personality may be terrific for a recreation director on a Disney cruise, but it is the mercurial, complex, challenging, and edgy personality that is more likely to transcend oddness and foster audacious creativity.

CREATIVITY RECONSIDERED: A SONG FOR THE COMPARISON GROUP

It is a commonplace of award ceremonies that the honorees deflect attention from themselves, the creative heroes, and thank the unsung support cast without whom the creative projects would never have been accomplished. Whether this is a genuine acknowledgment or a formulaic ritual is open to debate. But in this section I want to make an empirically informed case for acknowledging those who work with highly creative individuals. I want to focus on the comparison group in the IPAR studies and, in a sense, to sing a song of praise for the unsung.

Consider, first, the overall rationale of the IPAR studies. They were designed to identify the personality features of individuals who had creatively transformed their fields. In the case of the architect study, creative transformation entailed having completed architectural projects that were widely agreed to be innovative accomplishments that had an enduring impact. But consider for a moment how those creative accomplishments must have come about. Although creative individuals have many admirable qualities, they can also be a royal pain to work with. They can be self-absorbed, quick to anger, dismissive of detail work, and uninterested in the kind of social exchanges that make for a supportive and collegial working environment. So how on earth did those creative projects get accomplished?

There is a widespread assumption that the world's most audacious and innovative accomplishments arise from the mind of the single, isolated creative hero. This is a myth. I think we need to look more closely at the characteristics of the *comparison* group of architects. Remember some of their personality characteristics: they were responsible, sincere, reliable, dependable, clear thinking, tolerant, and understanding. They were also sociable, steady, and comfortable with detailed work. These are precisely the attributes needed for ensuring that a creative project is brought to fruition. The creative project—that which eventually reaches the outside world and transforms it—requires not only the innovator but also the contributions of the negotiator, the pacifier, the guy in accounting, the picker-upper, the soother who placates the hounds at the door, and the soft-spoken receptionist who tells you diplomatically that your fly is undone. It is true that creative heroes give greatly to others, sometimes in ways that dazzle. But they also are supported by other people with complementary personalities, without whom the innovative project would never get accomplished.

Given this interdependency of the creative and conventional, it is important to examine the possible costs that might be experienced by those architects who worked in the same firms as the creative stars compared with those who did not. The evidence from IPAR was that there may, indeed, be a price to be paid by those who work alongside creative individuals. There was a clear pattern of evidence that this group was lower in psychological adjustment, higher on anxiety, and more conflicted than comparison groups of architects who were not working with creative individuals. MacKinnon speculates that they are conflicted because even though they have many of the same attributes as the creative stars, they are unwilling to assert themselves to transform those potentials into action, whereas the creative stars are able to do this effectively.[15] It is easy to see how working alongside a creative person might create conflict and anxiety in individuals who have the potential to be creative stars themselves but are devoting much of their time and energy to supplying the collegial support that enables the creative stars to shine. I suspect something very similar happens in family dynamics and on sports teams.

CREATIVITY AND WELL-BEING: LESSONS FROM DARWIN

I asked at the outset whether you are creative yourself and whether you will still want to be after you've read this chapter. In many respects being a creative person and leading a creative life are demanding. First, there are the darker forces with which creative individuals struggle. Being open to experience means you are familiar with emotions like anxiety and depression that can spin out of control. Second, there is the constant pressure of going against convention—people do not relinquish their preferred and well-learned patterns of behavior without a fight. If you are creative, you will have experienced those looks of incredulity if not outright hostility that greet the true innovator. This can be taxing. And third, there is the sheer exhaustion that often accompanies creative work. Being passionately engaged in creative projects can interfere with sleep, cause tensions in relations with others, and compromise your physical health. Are you sure you want to be creative or continue being creative?

But there is a positive side of being creative. First, as we have seen, the negative emotions are there. But, being open, you are also disposed to experiencing positive emotions more keenly than conventional people. You readily feel joy, delight, and flow, and these can more than compensate for the bouts of negative emotion that openness entails. Second, although going against convention can be exhausting, it can also be elating when your creative project actually solves a problem that defies more conventional approaches. And that elation arises from the intrinsic motivation that impels creative pursuit; in fact, external inducements and recognition can actually be demotivating.[16] Third, although there are health costs that creative individuals might incur, there is a subtle twist when it comes to these costs that needs to be considered.

Consider the case of Charles Darwin. It is well known that Darwin suffered from an illness that kept him confined for years. He experienced recurring bouts of light-headedness, palpitations, vomiting, flatulence, and pain in the chest. The symptoms first appeared just before his famous five-year trip on the *HMS Beagle*, in which he gathered the information that would lead to his theory of evolution. During that trip Darwin was hardy, robust, and venturesome. He was symptom-free. But after his return to England the symptoms returned and, despite no evidence of a medical basis for the symptoms, eminent medical authorities strongly advised him to rest at home. There have been many attempts to explain the nature of Darwin's illness. One of the most interesting explanations is that of Sir George Pickering in his book, *Creative Malady*.[17] Pickering proposes that for highly creative individuals like Darwin, Florence Nightingale, and Marcel Proust, their various illnesses may have enhanced their creativity. Although physical illness would not be an effective ally for creative development, psychological illness might be. In the case of Darwin, Pickering concurs with an earlier diagnosis that the illness was psychoneurotic and that it had a function—to protect Darwin from the trivialities of social intercourse.

Darwin's letters provide ample documentation for this view. He turned down the invitation to be secretary of the Geological Society because it would have involved considerable social contact—"Of late anything which flurries me completely knocks me up afterwards and brings on a violent palpitation of the heart."[18] And it is clear that he was most flurried when confronted by social exchanges involving conflict. It is difficult to imagine a more controversial theory than that of Darwin's in an age that was still deeply conservative and in which people had fixed beliefs about creation. By becoming a reclusive invalid, Darwin traded off a life of social engagement for a life in which he could pursue his core project of writing up his emerging theory of evolution. Indeed, as Pickering documents in some detail, the creative writing of his theory of evolution was accomplished because everything else in Darwin's life was sacrificed to that monumental task. In later chapters I will talk about how well-being depends on the sustainable pursuit of such core projects, but for now it is important to allude back to our earlier discussion about the support that needs to be in place for audacious creative projects to be accomplished. It is highly unlikely that Darwin would have been able to sustain his creative work had it not been for the support of others, particularly his wife, Emma. Throughout their forty-three years of marriage she protected him from the intrusion of social stimulation, relaxed him with her daily piano playing, served as his secretary and editor, and, in later years, could be seen in the garden at Down Cottage with her beloved "Charlie."

We can conclude that highly innovative people, odd and audacious as they are, rely on the support of the unsung helpers like the IPAR comparison groups and Emma Darwin. Often this support of creative projects is not acknowledged, but in the case of Darwin, it is abundantly clear that he was constantly aware of how essential she was. He wrote of Emma: "She has been my greatest blessing. . . . She has been my wise advisor and cheerful comforter throughout life, which without her would have been during a very long period a miserable one from ill-health."[19]

By way of summary, what can we conclude about how research on personality and creativity can inform our reflections about ourselves and other people? If you scored high on Gough's creative personality scale at the beginning of the chapter, then you may have recognized yourself in the portrait we have drawn of highly creative people. You are likely to be open to experience and sensitive to sensations, images, and thoughts that others may ignore or may never even discern. This can, at times, be disturbing, both for yourself and those with whom you come in contact. But the products of your creative acts, the offspring— in Lady Gaga terms of "condomless" encounters with fecund ideas— may yield novel ways of solving problems, and these novel adaptations can benefit both you and others.

If you did not score high on Gough's scale, then I hope you have taken to heart the role that more conventional people play in the creative process. As the Machiavelli epigram notes, changing the order of things can be difficult, chancy, and dangerous. I believe that such challenges arise, in part, because the personality traits of creative innovators, though well suited for generating novelty, are not those best suited to bringing their creative idea to fruition.

Finally, it is important to realize that well-being has many different facets, and these facets may be in conflict. Your pursuit of creative ventures might bring inordinate satisfaction. It might become the defining cause of your life. It might change the world. But it might extract a toll on your health or relationships. So in the end it comes down to making choices about which of these various aspects of the good you accord the greatest significance. Follow your passion, by all means, but know that in so doing you might be choosing your poison as well.

chapter eight

Where Are You?
Personality in Place

WHERE ARE YOU RIGHT NOW? IN A DOWNTOWN COFFEE shop? At the bottom of the garden? Online? On that long, boring, noisy commute once again? Tucked away in a little reading nook that you have made all your own? Do you prefer pulsing, stimulating environments or those that are tranquil and serene? Are the places where you feel most yourself the places that make your partner squirm? Do you bare all in the Twitterverse, or do you regard social media as pernicious and threatening? In this chapter we examine these kinds of concerns about how our physical environments interact with our personalities to help shape and sustain our well-being. I will ask you to reflect on dragonflies, Times Square, Fargo, and Facebook. We will see that personality needs to be put in its place as we reflect on the quality of our lives. And as we shift focus from real places to virtual ones, from cities to Cyberia, the very nature of the concept of place itself will change radically.

DRAGONFLIES, HARMONY, AND WELL-BEING

I have frequently taught personality psychology to architecture and urban design classes and have found the students to be intriguing, challenging, and, to be candid, sometimes rather strange. Although my doctoral training was in personality psychology, I had a secondary interest in what was then the brand-new field of environmental psychology and at Berkeley had taken the first graduate class offering instruction in this field. During those years, the mid-1960s, architects and urban designers were keen to find out what psychologists had to say about the relations between people and places. And we, in turn, were intrigued about the tacit psychological assumptions they were using when they designed our dwellings and cities. So I scoured the architectural and design literatures and attended conferences where the environmental design fields and the behavioral sciences would converge. One of these meetings, held in Lawrence, Kansas, in 1975, was particularly exciting for me. That was because of Christopher Alexander.

Alexander was trained in both mathematics and architecture at Cambridge University and then was among the first graduates of the PhD program in architecture at Harvard. His *Notes on the Synthesis of Form* had a major impact on several fields.[1] It became a key text for the nascent field of software design and is still influential in a diversity of design fields. Its impact on architecture was more polarizing. In part this was because Alexander believed that the best designs for buildings arise not out of the creative architect's expertise but out of the timeless ways of building that are based on local knowledge. That we could essentially do without architects didn't go down that well with many architects. Alexander created what he called a "pattern language," a generative grammar of recurring environmental forms that evolved to satisfy human needs. To me this seemed a very promising way of looking at the linkages between people and places—of actually putting personality in its place. So when I heard that he would be giving the keynote address at the conference in Kansas I made sure I was in the front row with notepad at hand and a growing sense that I was about to hear something profound.

I wasn't disappointed. Alexander was tall and slender, rather like a reedy British Ichabod Crane. After being introduced, he stood very still for a few moments as though lost in thought, and then he started to speak, slowly and haltingly. His topic for the presentation, as I recall, was "What Is Architecture?" and he began with an image. He had been in Kyoto, sitting in a garden, when a dragonfly darted in from the blue sky and then, gently, landed on the petals of a cherry tree. "And that," said Alexander, pausing for effect, "is the essence of architecture." Then, another long silence.

I am not entirely sure what I felt at that point. Definitely intrigued but perhaps a bit confused. I leaned forward, eager to hear more. The person sitting next to me had a different reaction. He was a hard-nosed, tough-as-nails, rat-oriented, quantitative psychologist. He leaned over to me and said, "What the f— is he talking about?" Well, that sort of broke the spell for me. It brought home the reality that architects think differently from psychologists, or at least some psychologists and some architects. But what Christopher Alexander was getting at and what I want to explore in this chapter is how environments can be designed to advance human well-being. For Alexander, this could only be achieved if the linkage between creatures and their contexts was a harmonious one. This hardly seemed a contentious notion, but as we will see, it was and remains so.

On November 17, 1982, at the Harvard Graduate School of Design, these ideas were the focus of an extraordinary debate between Alexander and another distinguished architect, Peter Eisenman.[2] The debate is widely regarded as a classic, in part because of the barbed and rather obscene comments that were hurled about the room. Eisenman was a postmodern architect, a deconstructivist, who had worked with Jacques Derrida and was extremely well versed in the movement that sought to dethrone modernism and its emphasis on functional design. He believed that architecture should be challenging, whimsical, dissonant, and discomfiting. It should represent and then resolve chaotic tension. In short, it should reflect the anxieties and perplexities of the day, serving as a mirror of contemporary concerns and sensibilities. Alexander abhorred this approach to architecture. He believed that the design of dwellings and cities should provide a sense of harmony and congruence—like dragonflies on cherry blossom leaves.

Alexanderville: Intimate Connection and the Design of Cities

Design, for Alexander, should respond to the deepest needs of people who would live within the structures it creates. But if architects and designers are to create places based on psychological needs, what do they need to know?

In a remarkable chapter called "The City as a Mechanism for Sustaining Human Contact," Alexander addressed this question explicitly, drawing on an extensive array of psychological, sociological, and psychiatric research.[3] He proposed a universal human need for intimate contact that he regarded as essential for well-being: "An individual can be healthy and happy only when his life contains three or four intimate contacts. A society can be a healthy one only if each of its individual members has three or four intimate contacts at every stage of his existence."

These contacts needed to take a particular form. People must reveal themselves, informally, warts and all, without fear. This meant they need to meet nearly every day, and those exchanges need to be strictly informal, without any script or role shaping the interaction. The sole goal was to bare one's deepest sense of self to the other.

Alexander believed that prior to the Industrial Revolution small towns fully satisfied this need for intimate contact. But with increased industrialization people moved from more communal to more private dwellings that were set off from other houses. This led to what he called an autonomy-withdrawal syndrome, which posed serious threats to both individual and societal well-being. He saw it as a pathological belief in self-sufficiency and autonomy, perhaps symbolized most poignantly by the image of a child playing alone in a big garden, by herself. Many would see such an image as a positive one, but for Alexander it represented a system that had gone seriously awry and had become a threat to well-being, both individually and societally.

One solution to these threats to well-being could be achieved by a more psychologically informed design of housing so the need for intimate contact would not be frustrated. To this end, Alexander proposed a city design that would promote social contact, including such features as increasing young children's exposure to other children and increasing the likelihood of drop-in visiting among adults. His design was based on twelve geometric features that optimized the spontaneous coming together of people through high-density modular structures. I won't go into the details of the proposed design—let's call it Alexanderville—because a few years after publishing it Alexander felt that it was too constraining and too deterministic. Instead, what deserves our attention is his overarching view of how the design of cities could enhance the quality of our lives by meeting human needs.

As a personality psychologist I read this account of human need and environmental form with considerable interest and a certain amount of skepticism. Recall the truism with which we began Chapter 1: "Each person is in certain respects like all other people, like some other people, and like no other person." Alexander is proposing that all of us need the intimate interaction his cities are designed to promote. But, as we've seen repeatedly throughout this book, there are individual differences in personality—the ways we are like some other people and like no other person—that would very likely have made living in Alexanderville a delight for some, a matter of indifference to others, and, for some, hell on earth. The most relevant trait in this case is introversion-extraversion. Frequent, intense, daily interactions with three or four other individuals might well be the ideal environment for extraverts. But for introverts? I don't think so either.

Alexanderville, then, is a theoretical city designed to enhance our exposure to frequent, intense, informal exchanges—a high-density, stimulating place. Consider now, another view about cities, one that provides a rather different view about how urban stimulation shapes our well-being. Let's call it Milgramopolis.

Milgramopolis: The City as Overload

We first met Stanley Milgram in Chapter 1 when I was discussing the phenomenon of the familiar stranger. Milgram discussed familiar strangers in the context of a more comprehensive theory of how cities contribute to human well-being. Milgram's account of the city, at least with respect to its level of social stimulation, is diametrically opposed to Alexander's view. Milgram viewed the city as a source of stimulation, the cumulative effects of which had a decidedly adverse effect on human well-being.[4]

He proposed that an individual, when entering a city, is confronted with three demographic facts: large numbers of people, compressed space (and, therefore, high density), and social heterogeneity. The converging effect of these three factors is to create a psychological condition of "information input overload." Milgram argues that this overload is psychologically noxious and prompts people to use adaptive strategies that will reduce the amount and pace of stimulation from their environment. Although those stimulation-reducing strategies create a positive benefit at the individual level, they create problems at the societal level. Let's consider three of the adaptive strategies we can use for dealing with overload.

First, we can decrease both the quantity and quality of time we spend on these sources of stimulation. The differences between urban and rural settings in the tempo of everyday transactions nicely demonstrates this. The pace of life in cities is faster than in nonurban areas: people walk faster and interactions are shorter.[5] Striding swiftly means that often we simply don't see those people and events that might contribute to overload. And this applies as well to transactions. For example, one set of research studies looked at how long it takes to buy a stamp from a postal clerk in cities compared with smaller towns. City transactions were significantly faster. But it is likely also that in cities the quality of the exchange would be lower. In a small town a visit to the post office may well spin off into a discussion about the weather, your sister's "special friend" and the car parked outside her apartment last night, and whether your fashion choice of having matching outfits with your cat was really a good idea. In cities, although all these aspects of you might be equally deserving of comment, there is no time for any of it; there are five people behind you—"Have a nice day. Next!"

Second, we can disregard low-priority inputs. We can simply not attend to some of the stimulation coming at us in cities. It is easy to see how this adaptive strategy, though shielding individuals from overload, can lead to serious social costs. Recall the demographic fact of heterogeneity—cities have greater diversity, certainly of people, but also in terms of the types of events and situations one is likely to come across. In the case of people, one adaptive strategy that has social costs is to simply ignore those to whom you accord less significance. The possibilities are endless: you might filter out anyone over thirty, anyone under thirty, people with tattoos, short people, panhandlers, or anyone getting out of a Range Rover. Of course this means that whatever the filtering criterion is for you, it needs to be immediately discernable. So size, color, adornments and so forth are highly visible cues and, therefore, can serve as effective filters. But if your very lowest priority input were, say, postmodern sociologists of a certain political orientation, it isn't that easy to spot them in order to ignore them, although Birkenstocks, beards, and backpacks might offer a decent hint.

Third, we can block off input before it even has a chance of entering into our processing system. For example, people living in cities, on a per capita basis, are more likely to have unlisted telephone numbers (we'll get to cell phones later) than do those living in small towns. This is an effective way of cutting down on unwanted stimulation. But there is a more insidious and interesting way of staving off unwanted social stimulation: we can have unlisted *faces* that convey the message that its owner is not to be disturbed. I have even noticed what I believe are subtle differences in the way that women in particular employ the unlisted face technique in different cities. In Toronto, for example, it is a straight-ahead gaze with a slight trace of annoyance; in Montreal it is the same but with an added *je ne sais quoi* conveyed with slightly raised eyebrows, usually accompanied by high cheekbones. I have absolutely no published empirical evidence confirming these observations, however.

Now it would tax our cognitive resources if, upon entering a city, we needed to consciously decide to deploy these different ways of dealing with input overload and explicitly think about how we could reduce the demands and entreaties of others. But fortunately for us, if not for others, the task is made considerably easier because in cities, but not in small towns, we develop a norm of noninvolvement with others. This norm means that instead of having to explain why we don't intervene in others' lives, we are called upon to explain why we do. And this norm is extremely powerful, as witnessed firsthand by Stanley Milgram himself.

It began with Milgram's mother-in-law.[6] She had asked him why people were not getting up in the NYC subway to give their seats to elderly ladies with gray hair, a group with whom she clearly identified. Milgram, ever the curious researcher, resolved to find out. He recruited some of his students to volunteer to go into the subway in Manhattan and to ask people for their seats. He devised several different versions of this, but the most interesting one was as simple as this: "Would you please give me your seat?" Interestingly, many of the graduate students, after thinking about it, decided not to do it. However, one student eventually tried the daring deed, and the buzz back at the lab was "They're getting up!" But the assistant was finding it extremely stressful. Milgram then decided to find out for himself. He reported entering the subway, approaching a person, and almost choking on the request for the seat. He said he literally felt sick. What's going on here? The norm of noninvolvement is very powerful and so deeply internalized that flouting it extracts a toll. This, for Milgram, is the core experience of living in cities—they are generators of overload. We cope by adapting mechanisms for overload reduction. And then we make those strategies the *expected* way of behaving in cities so that civility itself becomes something for which we need to apologize.

Your Utopia, My Dystopia: You Want Us to Move There?

So let's stand back and see what we have learned by looking at cities through the lens of Alexanderville and Milgramopolis. For Alexander, cities enhance feelings of personal agency and isolation and should be radically redesigned to stimulate more human connection and intimate contact. For Milgram, cities have too much human contact, and such exposure leads to overload and the adaptive strategies that mitigate its negative effects.

Now it is true that Alexander is primarily concerned about environmental design at the level of houses and neighborhoods, whereas Milgram's focus is on the city center. And it is also true that Alexander is being explicitly prescriptive about how we ought to design cities for human flourishing, whereas Milgram's focus is on describing the experience of living in cities. But here's the issue: Alexander sees the city as a place where human contact ought to be amplified so our universal human need for intimate contact is satisfied, whereas Milgram sees the city as a place where human contact needs to be reduced so overload doesn't challenge our limited abilities to process information. Both assume that their perspectives apply to all of us. Both underplay the possibility that there are important individual differences in the need for stimulation, especially social stimulation.

Consider Alexander's initial depiction of human habitats as isolated, individualistic, low-stimulation domiciles. Such an environment is likely to be attractive to certain personalities—introverts and those who are high in locus of control, for example. And his proposed solution, what I call Alexanderville, with its high levels of social contact, is likely to be particularly attractive to agreeable extraverts and those who are open to experience. Conversely, Milgram's depiction of the city as a source of aversive overload, a frenzied sequence of "To whom it may concern" messages, is one that some people may actually seek out rather than avoid—extraverts, again, or perhaps especially Type A individuals.

In short, one person's utopia may be another's dystopia, and the design of our living spaces should, ideally, reflect what we know about how personality and place interact. And, as we'll see, we need to go beyond the Big Five to understand this.

ENVIRONMENTAL PERSONALITY:
EIGHT STANCES TOWARD THE ENVIRONMENT

Although the Big Five traits, such as extraversion and neuroticism, help us understand the kinds of places to which we are naturally attracted, these are rough guides only. Environmental psychologists have provided a much more refined set of traits, what they call environmental dispositions, to help us understand the full range of orientations toward our physical environments.[7] George McKechnie created the most comprehensive assessment device for tapping into these environmental dispositions, the Environmental Response Inventory (ERI).[8] The ERI provides scores on eight different dispositions toward our everyday physical environments. If you have been having some serious discussion with your partner, roommate, or family about the possibility of moving to a different city or town, you might find it interesting to think about the descriptions of the environmental orientations of individuals scoring high on each of the ERI scales. See if you recognize yourself in these descriptions.

PASTORALISM (PA)

People high on PA display sensitivity to pure environmental experience, opposition to land development, appreciation of open space, and preservation of natural resources. They also are accepting of natural forces as shapers of human life and endorse self-sufficiency in the natural environment.

URBANISM (UR)

People high on UR enjoy high-density living and appreciate the unusual and varied stimulation of urban areas. They take an interest in cultural life and enjoy the richness of human diversity.

ENVIRONMENTAL ADAPTATION (EA)

Those scoring high on EA regard the environment primarily as providing comfort, leisure, and the satisfaction of human needs, and they endorse modification of the environment to achieve those ends. They endorse private land use and the use of technology to solve problems, and they prefer stylized environmental details.

STIMULUS SEEKING (SS)

People high on SS express great interest in travel and the exploration of unusual places. They enjoy intense and complex physical sensations and display a great breadth of interests.

ENVIRONMENTAL TRUST (ET)

Those who score high on ET are responsive, trusting, and open to the environment and have a sense of competence in navigating the surroundings. They are relatively unconcerned about their security and are comfortable being alone and unprotected.

ANTIQUARIANISM (AN)

People high on AN enjoy antiques and historical places and have a preference for traditional vs. modern design. They have an aesthetic sensitivity to well-crafted environments, landscape, and the cultural artifacts of earlier eras. They have a tendency to collect objects for their emotional significance.

NEED FOR PRIVACY (NP)

Those high on NP have a strong need for physical isolation from stimuli and distraction. They enjoy solitude and dislike extensive contact with their neighbors.

MECHANICAL ORIENTATION (MO)

People high in MO are interested in how things work and in mechanics in its various forms. They enjoy working with their own hands and have an interest in technological processes and basic principles of science.

Consider now Donald and Rachel, a couple who are trying to make a decision about moving to a new residence. The decision isn't an economic one, let's assume, but a lifestyle choice, and they have the good fortune of being able to choose among a diverse range of desirable locations. Donald scores high on the ERI scales of Urbanism and Stimulus Seeking, whereas Rachel scores high on the scales of Pastoralism and Antiquarianism. They are unlikely to agree on a destination not simply because of "cold" intellectual differences but also because of the "hot" emotional stances they take toward their environment.

Consider Donald's high score on Urbanism, which, as McKechnie describes, leads to a value stance that

> the essence of human life lies in one's relationships with others. Cities bring together interesting and informed people, and sustain a cultural, aesthetic, and intellectual life that is impossible without dense urban clusters. Cities force interdependence among people, and this interdependence weaves together the fabric of human existence.

And his high score on Stimulus Seeking suggests he would also take this stance:

> Life is an adventure: there are things to do, mountains to conquer, cities to explore. To feel is to be alive, to be sensitive and responsive to the surrounding environment. The quest for adventure must not be inhibited by petty rules or conventions. The critical commitment in life is to the new, the unique, the untried, and the exciting.

It should not be surprising that Donald's preference is to move to a place where he can satisfy his desire for social contact, cultural diversity, and adventurous excitement. For Donald, the city trumps everything. Ideally, he wants a loft in the heart of the metropolis and to throw himself into the endless fascinations of urban life.

For Rachel, alas, the city is the least attractive of several options they are considering. Her high score on Pastoralism leads her to take a very different stance toward her environment:

> Appreciate the wonder and beauty of nature. Let it into your life, and let it shape your life. Be careful in all you do not to damage or squander the natural environment. Understand ecology and it will sustain you, for "in wildness is the preservation of the world."

Her high score on Antiquarianism suggests a more nuanced aspect of her preferences:

> Physical objects are the keys by which to unlock revealing memories and reminiscences. The gentle curve of a vase and the proud detail of a table give comfort, and provide emotional support, strength, and identity with which to face the future. Life is sustained through an emotionally charged and aesthetic closeness to and dependence upon the objects that define one's personal environment.

Rachel's preference is for a small village in the country where she would love to open a boutique store. She'll call it Silent Spring Antiques, selling fair trade coffee, consignment antiques, vintage clothing, and handmade furniture. She would actually like Donald to join her in the venture, but he has intimated to her that he would prefer to stick needles in his eyeballs.

For her part, Rachel truly hates big cities. Even if she has to go it alone, she is resolved to move to her rural retreat where she can have daily contact with curved vases, proud tables, and four cats. As for social stimulation, Rachel would be completely content with a small group of dedicated customers. They may not appear to Donald to be very exciting folks, but they could be relied upon to perk up and pitch in if a del Gesù violin mysteriously comes up for sale, a new shipment of Kopi Luwak coffee has been unpacked, or one of the cats goes missing.

In the final chapter we will give some further consideration to what happens when there is a discrepancy between the projects we wish to pursue and the places in which we wish to pursue them. Donald and Rachel will have a bit of negotiation to do.

WHO'S YOUR PLACE? ASSESSING THE PERSONALITY OF CITIES AND REGIONS

We have so far discussed environments in terms of their relatively objective characteristics such as their demography, the amount of stimulation they generate, and how well they provide access to social connection. But there is another aspect of environments nicely captured by Richard Florida's book *Who's Your City?*[9] From this perspective, places have personalities, so we might describe a city or town or neighborhood as extraverted, agreeable, neurotic, open, or, perhaps less obviously, conscientious. Jason Rentfrow at Cambridge and Sam Gosling at the University of Texas, Austin, have initiated a fascinating research program in which they have created maps plotting the geographic distribution of Big Five personality profiles of different cities and regions throughout North America and Great Britain.[10] The scores on the profiles are derived from respondents' average score on a massive survey (more than three-quarters of a million people sampled) that completed a Big Five Inventory online. The study also gathered data on important quality-of-life measures such as health, mortality, and social engagement. The results are intriguing not because of a number of stereotypes that were confirmed but because some of the results were rather unexpected.

Let's start with extraversion, the disposition to be outgoing, sociable, and upbeat. Which state has the highest concentration of extraverts? After seeing the results of the study I have posed this question to different audiences, and so far not one person has guessed correctly. The most frequent guesses were Texas, New York, or California. But these were all wrong. The most extraverted state in the Union is, in fact, North Dakota. Why? The researchers speculate that this may be due to the impact of migration from Chicago, which is at the hub of a circle of extraversion that may reflect the high proportion of sales people and other jobs involving a lot of social contact.

But I think there is another possibility. In 2008 North Dakota struck an oil bonanza centered in the northwest of the state. Although massive oil reserves had been discovered in the 1950s, it wasn't until new technology, fracking, was developed that it became commercially viable. During the period from 2005 to 2009 the oil workforce in North Dakota has gone from just over five thousand in 2005 to over eighteen thousand in 2009. Most of these workers are young males in specialized oil jobs, such as roughnecks, riggers, and roustabouts. They are ambitious, for the most part unconstrained by family ties, and almost certainly highly extraverted. Indeed, extraversion is one of the personality traits that has driven emigration since the first settlers left the predictability and comfort of their homes to explore new possibilities abroad. Extraverts go where the future seems most promising, and the oil fields of North Dakota have been no exception.[11]

What about the Big Five trait of agreeableness? This disposition to be pleasant and affable is high in the South. But the highest-scoring state of all was, once again, North Dakota. Is there some kind of Fargo factor or Bismarck bond that draws pleasant and outgoing people to North Dakota and keeps them there? We have already considered the financial inducement to head to North Dakota (or at least to western North Dakota), but although this seems clearly related to the trait of extraversion, it is not as clearly related to agreeableness. What attracts and retains people high in agreeableness are places known for small friendly towns where cooperation is the norm and conflict is minimized. Even moderate-sized cities like Fargo, which is a two-minute drive from Moorhead, Minnesota, are likely to be regarded as particularly friendly and pleasant. Indeed, the Fargo-Moorhead Conference and Visitor's Bureau captures this perfectly on the front page of their website, which has as its main heading, "A Warm Welcome Awaits You": "Forget the weather channel. They report degrees of heat. We're talking about degrees of warmth. When you get out and experience our community—and our people—you'll find that Fargo-Moorhead is one of the warmest metro areas in the nation."[12]

Another aspect of agreeableness is modesty, and in the warm welcome above, "one of the warmest" turns out to be a slight understatement. North Dakota and Minnesota are, respectively, the highest and second-highest states on agreeableness, and what is known as "Minnesota nice" is both a stereotypical and objective appraisal of the level of agreeableness that one experiences there. Here is an example: In 2004 a major flu crisis led to an exceptional rush on vaccinations throughout the United States and elsewhere, with long lines of people anxious to get immunized. Except, it happens, in the states of North Dakota and Minnesota—the center of the Nice Nexus. In one *New York Times* report it was noted that the residents were displaying their fabled agreeableness by foregoing vaccinations so others could have them during the crisis. The manager of the immunizations section of the Minnesota Department of Health urged the state to recognize that there were flu shots galore available for vulnerable people, but there were very few takers. "They call it Minnesota nice," she said. "People feel that they should defer to someone who needs it more."[13]

Although at the state level agreeableness is correlated with social engagement, religiosity, and civic mindedness, it is negatively related to the frequency of going to bars. Low scores on agreeableness are found to be most prevalent in the Northeast cities, to which they might well say, "I'll drink to that."

With respect to openness to experience, the disposition to be exploratory, curious, and creative, the Northeast dominates, particularly so New York City, where there is a disproportionately large number of people in the creative and artistic professions. This is consistent with what we know about the demography of creativity. New York attracts an extraordinary diversity of individuals who selectively migrate to where there is ample room to pursue audacious projects and other talented individuals who might support those pursuits. Did North Dakota score at the extreme on openness as well? Yes, indeed—dead last! Consistent with the strategy of matching individuals to environments with whom they are compatible, it might seem appropriate to recommend that agreeable extraverts pack up and fly to Fargo if they aren't there already. Except, if they fit the pattern perfectly, they are likely to be closed to the idea.

The trait of conscientiousness, which reflects a population that is dutiful, responsible, and self-disciplined, displayed a pattern that resembled that of agreeableness, finding its highest concentration in the southern US states and, rather contrary to stereotypes, its lowest scores in the Northeast. Perhaps the most surprising score was that of Florida, which, contrary to many of those pesky stereotypes, scores highest on conscientiousness. It is possible that this may partially reflect the large number of aging individuals in the state, because conscientiousness is reliably found to be higher among older individuals.

Finally, neuroticism also displays an intriguing spatial character that can be defined as a stress-belt roughly dividing the East and West. Neurotic places are characterized by emotional instability, anxiety, and impulsivity, and those who reside in such places have lower rates of exercise, higher disease rates, and shorter life expectancy. This cluster of characteristics was particularly prevalent in New York City. Where are the least neurotic places? They are in California, where the stereotype of mellow Left Coasters has much more than a shred of truth to it.

Personality psychologists are just beginning to explore these intriguing links between person and place. The significance of this work for the study of personality is that it highlights some of the sources of potential malaise in our daily lives as well as the joys of living in places that resonate with us. I expect that we will find strong pressures to migrate elsewhere on individuals who are living in cities that are discordant with their own personality. It is unlikely that someone transplanted to New York City who is affable, closed-minded, and devoid of any trace of neuroticism is likely to fare well in the Big Apple. Far better, perhaps, that he flies off to Fargo.

PERSONALITIES IN CYBERIA: CONNECTION RECONSIDERED

I asked at the beginning of the chapter whether you were currently online, and there is a very good chance that if you weren't then, you will have been sometime today. We are increasingly engaged with Cyberia, by which I mean the world of Twitter, iPhones, YouTube, Facebook, and countless other emerging technologies for connecting us all. As we become increasingly engaged with this world, as it becomes the environment in which we work, play, and express ourselves, now is the time to ask what effect it has on our well-being and how our Cyberian experiences both shape and are shaped by our personalities.

There are two conflicting views about Cyberia. One sees it in utopian terms as a means of efficiently and effectively connecting with others and exposing us to a limitless range of experiences and information. The other view is dystopian. It views living in Cyberia as overloading, artificial, and dehumanizing as well as drawing us away from genuine human encounter. In other words, from the optimistic utopian perspective, Cyberia is Alexanderville: it is a generator of human connection. From the pessimistic dystopian viewpoint Cyberia is a Milgramopolis—an overloading and taxing environment that leads to stress and disengagement. Some recent empirical studies cast light on each of these perspectives of our wired worlds.

Barry Wellman of the University of Toronto has carried out an extensive series of studies on the social connections the Internet and mobile technology promote. The theoretical framework for this research postulates that there is a new form of social organization occurring in the light of technological advances, what they call "connected individualism."[14] Instead of people being linked to others through groups, Wellman's group proposed that each individual forms a social network that has only partial overlap with the connections established by those with whom they live in the nonvirtual world. Given this new form of social organization, the challenging question was whether these kinds of networks provide the same kind of support and connection that more traditional forms of community provide. Some earlier studies appeared to show that engagement with Cyberia would draw attention and energy away from real social interaction with others, that the Internet and mobile technologies would have an isolating effect on individuals, with resulting costs in terms of stress and well-being. Contrary to those early concerns, Wellman's group have provided strong evidence that connected individualism is a positive development and that virtual connectivity enhances the quality of lives. For example, rather than drawing energy and attention away from engagement in the nonvirtual world, Internet connectivity facilitated greater, not less, real-world volunteering in an isolated community in northern Ontario.[15]

It is interesting to revisit Alexanderville in the light of research on connected individualism. Does virtual communication provide what Alexander postulated as the basic human need of intense, frequent, face-to-face contact, warts and all, with a close group of other individuals? For the aficionados of Facebook, Twitter, and other emerging social networking sites, the answer seems to be a resounding "yes." Now it is obvious that there are qualitative differences between virtual and real encounters. We can't smell each other on Facebook, for example . . . yet. But engagement with the Cyberian world seems to greatly enhance the essential function of easily communicating what we care about to valued others.

Wellman's group is sociological in orientation and not primarily concerned with examining individual differences in people's responses to Cyberia. But my students in the Social Ecology Research Group (SERG) at Cambridge University were particularly interested in exploring such individual differences, especially whether scores on personality traits and appraisals of personal projects could offer a glimpse into some of the more subtle relations between personality and the use of social media and connectivity. We focused primarily on Facebook and its various functions such as status updates, messages, chats, and wall postings. We were interested in exploring whether these ways of connecting with other people would enhance well-being, especially if such Facebook functions would enable people to make their personal projects visible to others and, therefore, enhance the likelihood that they would receive support for their pursuits.[16]

The results of these exploratory studies confirmed that although all users derived satisfaction from Facebook, they showed a preference for those functions that involved intimate exchanges with one other person (in most respects just like e-mail) rather than the more expansive functions visible to all, like wall postings and status updates. There were also interesting individual differences. Generally speaking, extraverts use Facebook more frequently and enjoyed it more, consistent with our earlier predictions about who would be most comfortable in an Alexanderville-type environment. The types of personal projects that the respondents made visible on Facebook were mainly those involving recreation, interpersonal, and academic projects. There were two types of personal projects that Facebook users seldom made visible to others. First, intrapersonal projects, such as those things you are trying to change about yourself, were seldom made visible, likely because they were seen as too intimate. And maintenance-type projects, like getting the car tires changed, were not made visible, probably because they are seen as unimportant. Those who occupy the Twitterverse, of course, have recalibrated what is to be regarded as important: many of us have "friends" who enlighten us regularly on their daily ritual of obsessive flossing or the preternatural excitement of watching the dog next door vomit.

There are also gender differences. Women were more willing to make their stressful projects visible to others on Facebook, whereas men were less so. These results are consistent with some of our earlier research, predating the rise of social media, showing that men who made their stressful projects visible to others in everyday life had lower well-being, whereas women doing the same thing had higher well-being. It seems as if making a stressful project visible for men enhances the stress, perhaps because it displays a potential weakness, whereas for women it reduces the stress because it opens them up to potential support.[17]

If the research of Wellman's group and SERG is consistent with the optimistic, utopic view of Cyberia, there is also research that is consistent with some of the concerns that the more pessimistic and dystopic perspective has expressed. Consider, for example, recent research on the rise of cyber-overload. The researchers were interested in determining whether the kind of overload associated with emerging technologies led to compromised well-being. In a two-phase study they determined that higher levels of cyber-based overload significantly predicted higher levels of stress and poorer health later in the study, even after controlling for all relevant demographic variables and baseline measures of health and stress. And, consistent with the argument we have been presenting in this chapter, they also found that personality had an important effect on this overload well-being link. Cyber-over-load was less likely to affect participants who scored high on sensation seeking, which is closely related to extraversion and to the ERI's stim-ulation seeking. They also found the same effect with what they call place-overload, which is basically the kind of overload we associated with Milgramopolis.[18]

In short, both the utopian view of Cyberia as a potentiator of human connectivity and as a source of unremitting overload, have seen empiri-cal support, although it is still too early to draw firm conclusions about the impact of new technologies on well-being. But our central point in this chapter, that relations between environments and well-being depend importantly on human personality, seems to be confirmed. We know that measurable aspects of personality and environmental dispo-sitions predispose us to liking and flourishing in certain environments. Some seek out the chaotic, unexpected, noisy, messy, exuberant places that beckon in big cities; others prefer the serene beauty of silence, tranquility, and isolation. For designers the goal is to create places that will accommodate the full reach of humanity—with all their individual differences in preference and personality—not just people who, like the planner or architect, have a particular sensibility. This is no small challenge for those who create our living environments. Many intrepid designers will reject such a goal outright, preferring instead to create universally loved places that enable all of us to flourish. But as psychol-ogists, we must remain skeptical. For every New Yorker who adores the unremitting excitement and chaos, there is another who aches to get away to a less taxing and more fulfilling place.

The rise of Cyberia and its suburbs—Twitter, Facebook, YouTube, and the rest—may provide one route out of the need for environments that are responsive to individual differences in personality and prefer-ences. Because we have a remarkable capacity to make our cyber-world infinitely idiosyncratic, we may find aspects of life in Cyberia more fulfilling than life outside it.

When the unrelenting stimulation of place-based and cyber-based overload overwhelms me, I can find a restorative niche on the web. I could pull up a video of a dragonfly alighting on a cherry tree in a garden in Kyoto and muse on the Alexandrian question of whether that too, in a compelling way, is the meaning of a new architecture. I suspect it is not, but to convince you, I need to turn to questions of what is natural and spontaneous and what is programmed and algorithmic. I need to explain how we invest ourselves in objects and pursuits, and, as I will show in the next chapter, I need to introduce you to the surprising insights we can gain by attending to the spit we are about to swallow.

chapter nine

Personal Projects:
The Happiness of Pursuit

IT WAS MY DAUGHTER'S TENTH BIRTHDAY PARTY, AND ABOUT half an hour before the party began she approached me with a strange request: Would I hypnotize the guests and turn them into various barnyard animals? Of course I declined for seventeen reasons, including ethical, legal, and practical ones (e.g., what if the cows started chewing on the chickens?). She was very disappointed but tried another tack: Would I do something the kids would find interesting—something "psychological"? Gulping, because ten-year-old girls can be more challenging than a room full of dyspeptic neuroscientists, I tried to think of something "interesting" to do for the delightful but daunting partygoers.

SPITTING IMAGES: GETTING PERSONAL

We gathered in the kitchen, and I asked for a volunteer. Jennifer volunteered (all ten-year-old girls in 1980 in North America were called Jennifer, except my daughter). I gave her the following instructions: "Okay, Jennifer, I want you to get some spit in your mouth." I demonstrated with exaggerated mouth movements, which they somehow found funny, and she followed by making her own good wad. "Now, swallow it." Jennifer complied, looking rather confused. I then asked her whether it felt strange or uncomfortable, and she responded that it didn't. Apparently she had swallowed on previous occasions. So far this was profoundly uninteresting, and I barely glanced at my daughter, who was showing incipient signs of mortification. But then I took out a sparkling clean glass, placed it in front of Jennifer, and asked her once again to get some spit in her mouth. She did. "Now spit into the glass." She did. "Now drink it."

Gross! No way! They all recoiled at the prospect of drinking their own spit. So I asked them *why* it was so gross compared with the normal act of swallowing. Another Jennifer had a very clever idea. She suggested that in normal swallowing the spit was warm, but in the glass it was cold, and this made it gross. So I suggested we heat up the glass of spit. Would it be easier to drink then? "Ewww," was the consensus sentiment. The little demonstration seemed to be sufficiently interesting that my daughter continued talking to me. And, as a bonus, I didn't need to clean up after any barnyard animals.

Why am I telling you this? I think it helps us understand, almost viscerally, the nature of what is truly personal and the subtle nature of selfhood. At what point does our own spit transmogrify from being warm and "me" to being cold and alien? Maybe just as it dribbles off our lower lip? Because the dynamics of spitting may be of substantive interest only for philosophically inclined dentists, I won't linger longer on it, although I will revisit it later in the chapter. What we will see, however, is that its lesson helps us understand the *personal* nature of the personal projects we pursue in our lives, to which we now turn.

PERSONAL PROJECTS: WHAT DO WE THINK WE'RE DOING?

The concept of personal projects has been introduced in earlier chapters, and I now want to expand that discussion.[1] Informally we can think of personal projects as the things we are doing or planning on doing in our everyday lives. Personal projects can range from routine acts (e.g., "put out the cat") to the overarching commitments of a lifetime (e.g., "liberate my people"). They may be solo pursuits or communal ventures, self-initiated or thrust upon us, deeply pleasurable or the bane of our existence. As our personal projects go, so does our sense of well-being. Explaining that linkage is what this chapter is about.[2]

Although personal projects are actions, not all actions are personal projects. Some deliberate actions may not rise to the level of the personal saliency that characterizes a personal project. "Personal saliency" refers to whether the action stands out as something significant for the individual. Also, personal projects typically extend beyond a momentary action; they are extended sets of action. They are also, importantly, action in *context*. This means that the interpretation of a personal project needs to take into account the contexts within which it is embedded. Take, for example, the personal project of "put out the cat." This might be seen as a relatively trivial pursuit, an almost reflexive act. And so it is for most of us whose lives are happily shaped by the coercive claims of calculating cats. But consider this context: you have severe arthritis and use a walker. But your house has four steep steps to the back door. You can only put the cat out by abandoning your walker, holding on to the railing as well as the squirming Mr. Kippy, and carefully negotiating the stairs to the outside door. This is no trivial pursuit; this is a personal project that takes skill, strength, perseverance, and a good sense of humor. Context matters.

In our research we have found that people typically report that they are pursuing about fifteen personal projects at any one time.[3] Obviously, people can't do every project simultaneously, the concept of multitasking notwithstanding. So managing personal projects requires skills, such as establishing priorities among the projects we pursue, reconciling conflicts between them, and making sure we do not become depleted because of their cumulative drain on our resources.

What Are You Up To? The Content and Categories
of Personal Projects

My students and I have examined the linkages between personal projects and well-being for a few decades now, and a clear picture is emerging of how project pursuit can promote or cause havoc with the quality of our lives. I have developed a method for studying these linkages, called Personal Projects Analysis (PPA), that allows us to examine the content, appraisal, dynamics, and impact of personal projects on your life. I see it as an alternative to traditional ways of looking at personality. Whereas traits might be said to inquire into those aspects of personality that you *have*, PPA inquires into those aspects of personality that you *do*.[4] We start by asking individuals to create a list of their current personal projects. We informally call this the "Project Dump": people just list, without any attempt to prioritize or overanalyze, the things they are currently doing or thinking of doing. You might find it interesting to do your own Project Dump right now.

So what are you up to? Over the years of asking this, I have found that some personal projects are extremely popular. The most frequently listed project is typically phrased as "lose weight" or something more specific, like "lose ten pounds." Conveniently forgetting the laws of conservation of energy, I sometimes worry that the trillions of tons that weight-control projects would liberate could affect the earth's trajectory. And it is interesting that the same result was found on the website called 43 Things (www.43things.com), which I encourage readers to explore. 43 Things provides a repository of personal goals that individuals are pursuing, together with feedback from people who have had experience with that particular goal and encouragement from those who find the goal to be worthwhile, estimable, or cool. In one of the website's analyses of the most frequently listed goals, losing weight was a clear winner. "Writing a book" and "stop procrastinating" have also come near the top on many occasions. It goes without saying that these personal projects represent the pursuits, primarily, of WEIRD people, a term that three University of British Columbia researchers created to characterize people living in Western, Educated, Industrialized, Rich, and Democratic countries.[5]

In contrast to the most frequently listed projects, some personal projects are singular, idiosyncratic pursuits, such as "Be a better Druid," or, one of my personal favorites, "Fill in the sinkhole in the backyard before Fred gets home." Again, context provides meaning for project pursuit. This particular project was launched when Fred, a rather elderly heart patient, was due to return home from the hospital after a serious operation. His wife didn't want him to think that the six-foot-deep hole in the garden was meant for the late him!

Sometimes the sequencing of personal projects is instructive. One thirty-year-old male listed the following projects:

Get my pilot's license.
Get a water bed.
Go to Colorado.
Go to the Bahamas.

And then, tellingly, he adds,

Get out of debt.

Other projects are notable for how they convey a glimpse of the individual's personality traits, such as "Tell my sister to get rid of her sickening boyfriend," followed by "Think, before saying something stupid." And it is easy to conjure up an image of the young woman who listed these personal projects: "Singing with friends," "Sitting back and enjoying the music," "Playing with my dog," "Hugging my friends," and, finally, "Try to be more easygoing." *More* easygoing? Seriously? I suspect if this person were to be any more laid-back, she would fall off altogether!

Personal Project Phrasing

The language we use when we are discussing our personal projects is an important determinant of whether they are likely to be successfully pursued. In a brilliant analysis of the linguistic features of personal project phrasing, Neil Chambers demonstrated that the way we phrase our pursuits has important consequences for how they will turn out as well as for our overall well-being.[6] He shows convincingly that personal projects phrased as direct action, such as "Lose ten pounds," are more likely to be successful and associated with higher levels of well-being than are those that are phrased more tentatively, such as "Try to lose weight." Indeed, Chambers suggests that "trying personalities," who are at risk for lowered well-being, should be encouraged to reformulate their projects as active doings. Don't think about possibly doing it. Don't just *try* to do it—just do it!

Categories of Personal Projects

Independent of the way we phrase our personal projects, the domain that projects are aimed at also has important implications for our well-being. The most frequently listed projects are interpersonal projects, work (or school) projects, health projects, and recreational projects. Although less frequent in number, what we call "intrapersonal" projects (e.g., "Be more outgoing," "Get control of my temper") are of particular interest. These are projects that are concerned with attempts to understand and change one's self. As we shall see later, such personal projects have an intriguing and somewhat paradoxical relation with our well-being. To understand this, we need to know not just what kinds of projects we are pursuing but also our appraisals of them.

HOW'S IT GOING? THE APPRAISAL OF PERSONAL PROJECTS

One of the important implications of emphasizing the *personal* nature of project pursuit is that different individuals may construe what is essentially the same project in radically different ways. Take the very common project of losing weight. For one person, say, an athlete, this may be part of a fitness regimen that involves various phases of bulking up and then tapering down in order to get optimal performance in her sport. It is an enjoyable and valued pursuit, it is self-expressive, she feels confident in her ability to accomplish her goal, and she has strong support from her fellow athletes. For the person beside her in the gym, though, "Lose weight" may be an endless source of frustration, anxiety, and stress. It may not really have been her project to begin with—she has succumbed to pressure from others, who have been judgmental about her lack of progress; she believes, based on past experiences, that she will gain back the weight she loses; or thinks losing weight is a prerequisite for her real project, which is finding a mate. Clearly "Lose weight" as a personal project has very different impacts on these two women's well-being. So an essential part of what we have studied in personal projects research is how people rate each of their personal projects on different appraisal dimensions. We have explored dozens of different dimensions and have consistently found evidence for five major factors underlying these appraisals: project meaning, manageability, connection, negative emotions, and positive emotions.[7]

Project Meaning: The Personal Significance of Our Pursuits

One of the most important features of personal projects is that they typically provide sources of meaning in a person's life. For example, on scales from 0 to 10, individuals appraise their projects, on average, to be consistent with their core values (7.7), important (7.5), self-expressive (6.8), absorbing (6.2), and enjoyable (6.1). When we examine the emotions experienced during pursuit of a project, the ratings on positive emotions are rated considerably higher than negative ones—for example, experiencing feelings of happiness (5.9) versus sadness (2.1). In short, for the most part the pursuit of personal projects is a happy one.

What kinds of projects are most likely to be experienced in a positive fashion—as truly meaningful? People are especially likely to appraise positively both interpersonal and recreational pursuits: love and leisure are clearly rewarding pursuits. Academic work for students and occupational projects for working individuals are consistently rated as less enjoyable and more onerous.

Self-Identity: Spit Revisited

One of the most intriguing dimensions of project meaning is that of self-identity—the extent to which you identify with a personal project and regard it as truly "you." In the first few years of research on personal projects we didn't include such a dimension in our PPA assessment device. But a telling exchange with a student in a night class convinced me we needed one. Shortly after having carried out some preliminary research with PPA, I had just finished my first lecture on personal projects. I was very keen to convey to the students what I felt was the importance of looking at project pursuit. I recalled how George Kelly (we met him in Chapter 1) used to finish his lecture series on Personal Construct Theory with the dramatic claim, "You *are* your personal constructs." So, rather presumptuously, I ended my lecture that night by saying, "You *are* your personal *projects*." A woman in the third row, arms crossed and face flushing, yelled out, "I am NOT my personal projects." "Or *not*, as the case may be," I concluded, somewhat abashed. As the class left the auditorium I went up to the student, whom I knew well, and asked her why she was so piqued. She was an outstanding student, rather older than most, who had left university to get married and now had young children and was trying to complete a degree while working in public service. She told me that when she listed her projects in the Project Dump, all of them, except attending this night class, were someone else's projects—externally imposed tasks she felt obliged to undertake but didn't reflect who she was or who she might become.

Driving home that night I reflected on how powerful the sense of self-identity might be: some projects are warm and natural; others are cold and alienating. My mind wandered back to the little spit demonstration I'd performed four years earlier at the birthday party that had demonstrated the compelling claim of the bodily "me." Personal projects that are high in self-identity might also have a subtle and powerful impact on the person pursuing them.

When I got home I was rather tired from the evening lecture and went downstairs to watch my favorite sports show and chill out. I was pleased to see there was a bowl of peanuts on the coffee table and began to eat them. They weren't very good. On my way back upstairs I said to my daughter, "Those were the blandest peanuts I've ever eaten." She looked horrified and said, "Oh Dad, you didn't eat those, did you?" I was rather puzzled until she explained to me what had happened. Apparently she was on the kind of weight-control project that many young teenagers engage in, and instead of eating the peanuts in the bowl, she had just sucked the salt off each of them and put them back in the bowl. I had eaten a bowlful of presucked peanuts. Gross!

Self-Identity in Adolescents' Personal Projects

From that point on I was convinced that there was some intimate link between self-identity and the phenomenology of saliva in project pursuit. I imagined writing up a grant proposal called "Great Expectorations" and was fortunately able to suppress that idea, which partially explains why I actually did get a grant to study self-identity in high school students. I was able to assess each student in a large high school on the content and appraisal of their personal projects. Imagine what kinds of personal projects are regarded as most personally meaningful and self-defining for high school students—or, if you wish, what kind of projects are warm and "them" rather than cold and "other." Here are the highest average scores in ascending order, in which the possible range was from 0 to 10:[8]

Sports	8.2
Boy/girlfriend	8.5
Sex	8.6
Spiritual	8.9
Community	9.8

A few things are noteworthy about these results. First, given the relatively high frequency of explicit sex projects these students listed, we thought it appropriate to differentiate them from those simply listed as projects involving boyfriends and girlfriends. That they would be among the most enjoyable projects was expected and confirmed, but they were also those projects within which they felt most themselves. I found that intriguing because for years I had lectured on theories of human development in which it was postulated that individuals had to have a sense of personal identity *before* they could move on to intimacy with others. Our results suggested that, at least from the perspective of the personal projects pursued by students about to enter into the stage of emerging adulthood, identity and intimacy were not separable but coconstituted: we develop a sense of who we are by discovering how we are with intimate others. Further, when I have presented these results at conferences, attendees usually express surprise about the two most deeply self-expressive projects: spiritual projects and, especially, community projects. I should point out that these are not high-frequency projects—there are relatively few of them listed in high school students' Project Dumps—but for those who pursue them, they are felt to be deeply self-expressive.

Is it possible to find something common to each of these categories of highly meaningful personal projects? One possibility is that these are all pursuits in which students learn about giving and receiving from others, of being needed, and of developing a sense of competency in establishing intimate links. Sports might seem a bit of stretch in this respect, but most of those listing sports projects were boys, and team sports in particular might give them an opportunity to form extremely close bonds with others.

If we look at the other end of the self-identity spectrum, at those projects that are least self-defining, the results are rather disconcerting. Here are the lowest-scoring categories in descending order:

Reading	6.2
Maintenance	6.0
Academic	5.7

Regarding reading, it has to be said that these data were collected before the *Harry Potter* Revolution. I strongly suspect that if we were to do this study today, the reading projects would have a higher rating than what we found then. Maintenance projects are primarily things like cleaning a room or cutting the lawn, and their low level of self-identity probably arises in large part because they are likely imposed upon students rather than spontaneously sought out. I think the most dispiriting finding is that regarding academic projects. At least for these students, the academic pursuits in which they are engaged are the least self-expressive of any of the major project domains we examined; indeed, the scores on the last two categories, maintenance and academic projects, suggest that two of the most alienating things you could tell your teenager are "Clean up your room," and "Do your homework." It's almost like asking them to drink a saliva spritzer.

HOW ARE YOU MANAGING? INITIATION, EFFICACY, AND CONTROL

Let's assume you are pursuing personal projects that are deeply meaningful for you. You identify with them, they're consistent with your values, and they offer sufficient pleasure that you are motivated to pursue them enthusiastically. But how easy are they to manage? Is it possible that merely having meaningful pursuits may not, in itself, be conducive to well-being? Three of the dimensions we examine in PPA tap into the extent to which your personal projects are effectively organized and moving along well. The first, *initiation*, asks whether you were the primary initiator (high scores) or, as in the case of the mature student we discussed earlier, if it was other people (or cats) who initiated the project (low scores). *Efficacy* assesses whether you anticipate your projects to have successful outcomes. High scores (in the 8 to 10 range) mean that you see them as being highly likely to succeed; low scores (below 5) mean they are almost certainly not going to succeed. *Control*, as we saw in Chapter 5, is a key psychological variable that has important links with success. But whereas in the earlier chapter we looked at control as a relatively stable trait, here we see it as characterizing the particular projects you are pursuing in your life. These three appraisal dimensions, just like the meaning dimensions, all tilt in the positive direction on the 0 to 10 scale. Most of us initiate our own projects (7.1) and see them as likely to succeed (7.2) and as being under our personal control (7.3).

The *initiation* dimension is particularly interesting in light of a cross-cultural study I carried out with one of my graduate students, Beiling Xiao, on the personal projects of university students in China. We were interested in seeing how the content and appraisal of their projects compared with a matched group of North American counterparts, and we expected to find that North Americans would have a higher sense of self-initiated projects than would students in a more collectivist culture. My lab was very excited when the first translated results came through, and one of the projects listed caught our attention. It was simply, "Fix my *guilt*." We puzzled over whether this might have something to do with the reestablishment of churches in China, with a corresponding increase of experiencing a particular kind of guilt. We went back and forth on this theme for some time. But it seemed to be worded a bit strangely, so I decided we should go back and look at the transcription and recording of the data. I'm glad we did. It turned out that the handwriting on the translation had been misread. It turns out that the project was actually "Fix my *quilt*." Big difference! But we also noticed that this had been appraised as a project that was low in initiation, which we found a bit puzzling. We then noticed that other projects were consistently rated far lower in initiation than we had obtained with other groups of students. We ran the stats on this, and it confirmed our suspicions: Chinese students were, by a large margin, far less likely to initiate their projects than were their counterparts in Western countries. With a little probing we determined that this was likely due to the strong influence that the cadre or group to which an individual belonged exerted on the initiating of everyday projects. This seemed consistent with the communitarian—indeed, communistic—society within which the students' daily projects were embedded. This underscores an important aspect of our personal projects: they reflect not only our basic needs and personalities, but also, both in their content and their appraisal, the places and the political contexts in which we live our lives.

The *efficacy* dimension is assessed by asking respondents to rate the degree of progress on their personal projects as well as their likelihood of success. We have found consistently that this appraisal dimension is the best positive predictor of well-being, and the result holds across a broad array of samples and ages.[9] These results are consistent with the considerable literature in cognitive behavior therapy showing that efficacy is a powerful determinant of an individual's ability to cope with a diversity of problem behaviors.

Remember earlier when I asked whether simply having meaningful personal projects was sufficient to enhance well-being? Surprisingly, the answer is "no." Having highly meaningful pursuits does contribute to well-being but only rather modestly.[10] Now we can ask the same thing about having manageable projects. Is being engaged in manageable, efficacious projects sufficient to enhance well-being? Take a look at the personal projects you listed, and ask whether these are highly meaningful and also likely to be successfully accomplished. Our research suggests that whether they are accomplishable is more likely to enhance your well-being than whether they are meaningful. Indeed, we have postulated that there might be a meaning-manageability trade-off in project pursuit, such that your most meaningful projects are likely to be the most challenging in terms of their day-to-day management. It seems counterintuitive to claim that the pursuit of projects like "Take out the garbage," "Pick up the mail," and "Buy some tooth-paste" would be more conducive to well-being than "Grow as a person" or "Transform Western thought." So what is going on here? The best answer, I believe, is that well-being is enhanced when both efficacy and meaning are experienced within the same projects. In other words, mere efficacy is insufficient.[11]

The *control* dimension is very similar to that of *efficacy* in that both are concerned with the extent to which you feel you can influence the events in your life. We saw in the chapter on control that a sense of perceived control is an important determinant of well-being, but in that case we were investigating the concept of control as a gener-alized, trait-like disposition. Now we are concerned with how one perceives control over one's current and anticipated projects. Some personal projects are almost certainly under one's control, particu-larly those of the "feed the cat" variety, even though undue finicki-ness or ferocity (of the cat) might occasionally challenge that illusion. But projects like "Help Dad understand what's happening to him" or "Mobilize the resistance movement" may be impossible to control, despite our love and our resolve. The vicissitudes of life can conspire, as we've seen before, to wrest control from us and shake up our lives, and when our illusions are shattered, the results can be serious, both physically and psychologically.

And so it is with our personal projects. The formal way we put this in our theoretical work is this: *a sense of control is adaptive to the ex-tent that it is based on an accurate reading of ecosystem resources and con-straints.* This goes back to the issue we discussed earlier, of making sure our buttons are hooked up when we gear up for new pursuits. At least for some of our personal projects, taking stock of the resources we have available for helping us, including other people, is important. So too is it important to take stock of some of the barriers that might be expected along the way, including other people. Sometimes it is difficult to know just how and when those facilitators and constraints will play out. Here is where it is helpful to seek advice from those who have pursued simi-lar personal projects. On the 43 Things website, where people list their aspirations and goals, one of the most valuable features is the feedback others give. The value of this kind of feedback can be considerable given the well-documented difficulties most people have in predicting how they will feel about events in their personal future. Dan Gilbert in particular has provided extensive and compelling evidence that most people are unable to successfully predict their own future happiness, and the experiences of others who have already pursued the same options can better guide them when they are considering life options.[12]

WHO CARES? SHARING, SUPPORTING, AND CLEARING THE DECK

It matters then that our personal projects are meaningful and manageable. But what if other people think your pursuits are useless, misconceived, weird, or deeply evil? How important is a sense of support in your project pursuits? Or, more broadly, how are our projects connected to other people?[13]

In personal relationships it helps if your partner values your projects and vice versa. Having a partner who signals disinterest in or disdain for your favorite pursuits is dispiriting. Indeed, Anne Hwang, in her Harvard dissertation, showed that the best predictor of relational satisfaction among young adults was the extent to which they *shared* their personal projects.[14]

To connect our projects to other people, we need to make others aware of them by making them visible (or audible). Some individuals wear their personal projects on their sleeves, whereas others carry them deeply within themselves. As discussed, there are gender differences in how we deal with project visibility. With stressful projects women benefit from making visible the projects and their challenges in pursuing them, whereas men benefit from keeping that to themselves. This is likely related to gender differences in coping styles, in which men are primarily geared up to fight (or flee) when confronted with a stressful situation, and women are more likely to bond with others in order to meet the challenge.[15]

We observed something similar when asking men and women who were in senior management positions about the organizational factors that contributed most to their well-being.[16] For women the most important factor was the extent to which their organizational cultures seemed supportive of their projects. For men the most important factor was the extent to which the organization allowed them to pursue their projects without impedance, one in which no barriers blocked their project pursuit. For them the best support came from those people who knew when to simply clear the deck.

Unlike a sense of efficacy, which has a strong and direct relationship to well-being, social connection has a more nuanced and specific role to play. Two studies illustrate this very nicely. In one we studied women from the early days of pregnancy to their experiences of delivery. By looking at "pregnancy as a personal project," we were able to get the expectant mothers' rating on various dimensions and then correlate them with both subjective and objective measures of successful delivery.[17] The dimension that best predicted both kinds of successful outcome was the emotional support of their partners. More recently Craig Dowden presented some compelling data on entrepreneurs, looking at the dimensions of their personal projects that best predicted success both in terms of subjective well-being and financial success.[18] What was the best predictor? The emotional support of their partners. In short, when entrepreneurs talk about a particular project as being "their baby," the obstetric metaphor is a compelling one.

HOW DO YOU FEEL? THE EMOTIONAL LIFE OF PERSONAL PROJECTS

Let's summarize what we know about personal projects and well-being so far: well-being is enhanced if your projects are meaningful, manageable, and effectively connected with others. But what if, despite these favorable characteristics, your projects are utterly joyless and unremittingly stressful? Caring for a parent with dementia is an increasingly prevalent example of such a project, one that can bring a whole family to its knees. And as for positive emotions? To be engaged in pursuits in which we feel joyful and fully alive has a major influence on the quality of our lives.

Consider first the negative aspects of project pursuit. We have consistently found that well-being is strongly associated with the absence of stress and negative emotions experienced in project pursuit. Whereas efficacy is the strongest positive predictor of well-being, stress is the strongest negative predictor and at about the same level of magnitude. Let's put this in perspective: simply knowing whether the personal projects a person is pursuing are stressful predicts differences in well-being well beyond knowing that person's socioeconomic status, race, gender, and other key demographic factors. And if we look at the obverse of a life going well, one in which the individual expresses high levels of overall negative affect, particularly depressive affect, the same findings hold in reverse: depressed individuals are engaged in stressful projects that are low in efficacy.

Might there be differences in how the emotional life of personal project pursuit plays out in different cultures? Much remains to be learned, but we did explore this in one study that compared the emotional experiences reported by Canadians and Portuguese in their everyday personal projects. I had a personal interest in this particular comparison because my son married into a Portuguese family, and we became intrigued with some of the differences we had been told of regarding the expression of emotions in the two cultures. I was particularly interested in fado music and its association with the emotion of *saudade*, which has frequently been described as one of the most difficult emotion terms to translate into English. I knew that saudade was a kind of nostalgic longing, and after our Coimbra colleagues had treated me and my wife to an evening of fado music in an underground taverna I was even more intrigued. I decided it was time to get some local knowledge about saudade and get some exemplars for how it was used in everyday life. So when we went into a bookstore in Porto the next week I started to chat with a young man in the English-language section who looked like a graduate student. I told him that I was very interested in studying human emotions and wondered whether he could tell me what saudade was. His English was very good, and he had a dramatic flair about him. He paused and thought before responding, "Think of your wife going faaar, faaar away for a looong, looong time"—he glanced back and forth at us both—"What would you feel?" I knew exactly how I would feel, but my inner imp occasionally slips out, and I repeated his question and said, "What would I feel? Relieved!" Fortunately my wife knows I adore her, and after I assured the student, repeatedly, that if Susan and I were separated, I would be sad and anxious and FULL OF SAUDADE, he eventually left us. Soon after this exchange my Coimbra colleagues, Margarida Pedrosa De Lima and Isabel Albuquerque, began to gather data with me that looked at how positive and negative emotions play out in everyday personal projects. In comparing Canadian with Portuguese appraisals we found that the Portuguese experienced more positive emotions in their projects. They gave significantly higher appraisals to such feelings as hope and happiness and a considerably higher rating on feelings of love experienced in daily projects. But they also scored significantly higher on feelings of depression. And they scored significantly higher as well on feelings of emotional ambivalence in their pursuits. This combination of love, depression, and ambivalence seems to me to capture the essence of saudade, and it suggests the possibility that this might be a broader emotional stance characteristic of this particular culture and not restricted to romantic pursuits.

Let's assume then that your personal projects are meaningful, manageable, socially connected, and involve a higher ratio of positive emotions to negative emotions. Our research provides ample evidence that if your life is filled with such projects, you are likely to be happy and that, for you, life is good. And what if the opposite is the case? What if your days are played out in projects that have no significance, are chaotic, are isolated from the recognition and support of others, and are unremittingly painful emotionally? What then?

Unlike factors we have treated earlier in the book, such as relatively fixed traits and constraining environments, personal projects are tractable—we can change them. Whereas traits are something we have, projects are something we do. Whereas contexts embed us, projects pull us forward into new possibilities. And one of those possibilities is a better life and a happier life.

In the final chapter we will see how such an outcome can be possible. It requires that we understand the nature of *core* projects in our lives and how the sustainable pursuit of such projects is the key to our well-being. It will also mean that changing our personal projects may take us a step beyond the conventional, the warm, and the comfortable to something that makes us rather uncomfortable at first, even though it arises from our very selves. It means examining your deepest aspirations objectively, revising them appropriately, and then reincorporating them back into your core self. It means swallowing hard and making yourself vulnerable while changing your life. That takes courage, and it is definitely nothing to spit at.

chapter ten

Self-Reflections:
The Art of Well-Being

I AM NOW IN A STAGE OF MY ACADEMIC CAREER THAT FALLS somewhere between Full Professor and Incipient Senility. I call it my "Anecdotage." One of its symptoms is the irrepressible tendency to tell stories, occasionally even relevant ones, in order to clarify a concept or to stay awake when I'm delivering my lectures. So, perhaps predictably, I'd like to tell you an anecdote about telling anecdotes. But I promise my meta-anecdote will be deeply relevant to this concluding chapter.

I had been on a panel organized by the fine Educational Development Center at Carleton University, where several of us were discussing the pleasures and perils of professing. During the Q&A session a young chemistry professor asked us all a simple question: "Any thoughts about doing our last lectures?" Ah, last lectures! I couldn't wait to answer that one. I told him how I heard that the University of Michigan had institutionalized the importance of last lectures in its Golden Apple Award for distinguished teaching. Winners are asked to prepare and deliver their "ideal last lecture." This intriguing framing of the lecture had been inspired by the words of a second-century rabbinical sage, Eliezer ben Hurkanos, who had admonished his students to "Get your life in order one day before you die." As most of us do not know what day that will be, we need to get it in order every day. The Golden Apple award adopted this notion to celebrate those professors who consistently teach *each* lecture as if it were their last and who not only disseminate knowledge but also engage and inspire students in its pursuit.[1]

As I was relating this story I noticed that the chemistry professor was looking rather puzzled, and I suspected this could have been because I had totally missed his point; indeed, I had. "Brian, I was only asking about the last class we give during a lecture course, not the final one of our lives! Do you do a review? Do you say what's going to be on the exam? Do you tell them where to pick up their lab reports? That sort of thing." I had been blathering on about the poetry of professing, and he had been asking about the plumbing.

But when it comes to final lectures or, in the immediate case, a concluding chapter, I think plumbing and poetry both have a place. So what follows will be a combination of nuts, bolts, lyrics, and invocations. I will review the key concepts from previous chapters and suggest some emerging themes that tie things together. One theme in particular will be highlighted: how the sustainable pursuit of core projects enhances our well-being. Examining sustainable pursuit helps us reflect on the way our lives have gone and provides a perspective on the viability of our possible selves and personal futures. But then, as we pack up to leave, I want to pause and have a more intimate word with you about the deeper significance of what we have been pursuing throughout the book. I would like you to take this brief concluding conversation very personally.

THE SUSTAINABLE PURSUIT OF CORE PROJECTS

Chapter 9 discussed the importance of both the content of our personal projects and our appraisal of our progress or success in enacting or accomplishing those projects in determining the quality of our lives. I now want to sharpen this to propose that the *sustainable* pursuit of our *core* projects shapes the quality of our lives—our health and happiness and well-being broadly defined. We'll start first with the notion of core projects and then examine the diverse factors that influence their successful and sustainable pursuit.

Getting to the Core: Linkage and Resistance

Some personal projects become self-defining commitments of our lives and provide a deep sense of meaning for us. I call these *core projects*. How can you tell which of your personal projects are truly *core* projects? There are several ways of assessing this. First, we can identify those projects that you rate as most meaningful to you, on bases such as their importance, their consistency with your values, and whether they are self-expressive for you. Those projects that are meaningful in each of these different aspects can be regarded as core projects.

Another way of looking at core projects is to determine how each of your projects relates to the other projects that you are currently pursuing—in other words, by looking at your personal project system as a whole. Some projects are tightly connected with others in your system. If you are doing well on them, you are doing well on the others. If you are in difficulties with them, the rest of your project system is at risk. Shake a tightly linked core project, and the rest of your undertakings rattle as well.

Consider two individuals, each with a project of "Write a book," an aspiration that is consistently among the most frequent of the goals listed by individuals contributing to the website 43 Things. For one person, writing a book may be a rather peripheral project. She perceives it as having insignificant impact on her other projects, either positively or negatively. It is just something she decides to do because it seemed appropriate and worth doing. It would be cool to accomplish it, but it isn't a self-defining venture. It isn't expressive of her deepest values—she is more concerned with health issues and the well-being of her children than with self-expression in a book. For another person, however, writing a book may be intimately linked to all the other projects she is pursuing. Writing a book, in her view, will bring both cash and cachet and will increase her chances of meeting important people. It will also go a long way toward providing proof to her mother-in-law that she is something more than an inconvenient drag on her awesome son. And even more importantly, it will keep her focused and fulfilled like nothing else. It is a personal trademark for her at this stage of her life. At a superficial level the two women have the same project of "Write a book," but they are very different *personal* projects: one is peripheral and optional; the other is core and matters dearly.

An important consequence of the centrality of core projects is that they are not likely to be relinquished, even when there is a good reason to do so. Abandoning a core project would entail a major shift in the rest of the commitments and undertakings of your life. Resistance to abandoning core projects also means that you are not easily swayed by competing projects or alternative opportunities that knock on your door late on a Wednesday afternoon. But resistance can take a toll as well, particularly if a core project is pursued with grim resolve despite having lost its motivational force and its viability. Under such conditions the project becomes unsustainable, and your quality of life becomes compromised.[2]

So let's dig deeper into how we can enhance the sustainable pursuit of core projects. Three emerging themes drawn from preceding chapters provide both a summary of where we have been and a vantage point for reflecting on where we are going in our lives and our capacity to thrive. The three themes are essentially strategies for sustainable pursuit: adaptive reconstruing, self-change, and context monitoring.

Adaptive Reconstruing: Think Again

In the first two chapters we discussed the advantages of construing our worlds in a complex and adaptable fashion. As we saw in Chapter 1, when we meet a new acquaintance it helps to go beyond first blushes and to view the person's behavior through a more elaborate set of goggles. Although first blushes might be fine for short-term, inconsequential interactions, they could be wildly misleading when evaluating a prospective romantic partner or a potential business partner. Flexible and adaptive construing, in short, gives us more degrees of freedom in which to shape our actions and engage with our environments. Later we saw how creative individuals have an openness to experience that fosters complex thinking, including the ability to hold, simultaneously, conflicting views about events and objects. Let's examine now how a similar ability to adopt complex, adaptable, and flexible ways of construing can help sustain our core projects.

When we launch new projects they are fresh and meaningful, and more often than not, we approach them with a sense of efficacy and optimism. But over time projects, even core projects, can lose their luster, become increasingly incoherent and appearing disconnected to the changing contexts of our lives. When projects go stale in this way it doesn't augur well for their sustainable pursuit, and the quality of our lives can then suffer. But might it be possible to renew them by reconstruing or reframing them? Here are two good examples of how reframing our projects can lead to positive outcomes.

The first example involves the personal reflections of two distinguished organizational psychologists, Karl Weick and Jane Dutton, colleagues and friends at the Ross School of Business at the University of Michigan.[3] They each published chapters in a book on renewal in professional lives and provided a candid and touching portrayal of their contrasting ways of accomplishing this themselves. Jane talks about gardening, a passion of hers, and how it serves as a rich metaphorical base for reinvigorating her professional projects. Karl, in an e-mail exchange with Jane, shows how he approaches renewal of his projects very differently. Karl agrees that gardening works well as a metaphor for project renewal, and he is particularly interested in how Jane needs to weed her garden and her projects so her core projects may more readily bloom. But Karl looks at it somewhat differently: "You work on longer projects (e.g., 6 years) than I do. You think of renewing a life. I think of *moments* of renewal, which occur more often." Karl also notes another difference: "You weed in order to sustain large projects. I shrink projects in order to accept more of them." Finally he observes that the content of their metaphorical gardens is very different: "Your garden is full of people. My garden is full of books. Your relations are face to face. Mine are vicarious."

That these differences are related to broader differences in their overall orientation to their projects is made clear when Karl recounted a time when both of them presented papers at a conference. Jane wore some new eyeglasses with a faint red tinge, whereas Karl was wearing reading glasses. Her glasses revealed the audience in loving color; his glasses blurred his audience and sharpened the image of the paper he was reading. She literally looked at the world through rose-colored glasses, whereas he viewed it through highly focused, cooler lenses. But in reflecting on their different ways of seeing the world, he affirms that both perspectives have value.

Hotel room attendants in Boston hotels provide a second example of how project reframing can enhance our well-being, as demonstrated in research by Alia Crum and Ellen Langer at Harvard.[4] The attendants typically clean fifteen rooms a day, each taking about twenty to thirty minutes. It's a tough, repetitive job. But many room attendants complained that they got little physical exercise—or so they thought— and many experienced burnout quite quickly. Crum and Langer were interested in what would happen if the attendants were made aware of the potential health benefits of their daily routines: Might such awareness actually have a placebo-like effect that could be detected with physiological measures? The attendants were randomly assigned to two groups. One group was informed that cleaning rooms is healthy exercise and satisfies the Surgeon General's recommendations for an active lifestyle. Participants in the other group were not given this information. Four weeks after the intervention the group that was offered the "it's good exercise" reframing of their everyday cleaning projects showed a decrease in weight, blood pressure, body fat, waist-to-hip ratio, and body mass index. In short, project reframing—adopting new goggles for construing what you think you are doing—can have healthy consequences.

Personal Metaphors and Project Reframing

Projects can also be reframed through the strategic use of metaphor. I have used an approach drawing on metaphors to help managers and other professional groups creatively reframe personal projects that have become stuck and dysfunctional. The idea is to use a person's or a group's specialized domains of expertise from which to draw a rich array of associations and then apply this to the projects that need remedial attention. Specifically I ask people to create two lists. The first consists of elements or aspects of the stuck project, and the second consists of elements of the metaphor domain. Once these two lists are written down we then see how the metaphor might provide ideas for unsticking the problematic project.[5]

In doing this with senior military officers one participant, let's call him Colonel Poutine, identified one of his stuck projects as "Deal with the lack of morale of my junior officers" and listed as illustrative elements of that problematic project "laziness," "lack of follow-through," and "friction with fellow officers." After considering other domains that his fellow officers nominated—knitting, Thai cooking, fly fishing, and the art of seduction—Colonel Poutine chose ice hockey as his specialized domain (he was, after all, Canadian) and listed as some key elements "goal," "off-side," "assist," and "penalty shot."

The next step involves scanning the two lists to identify any possible connection or links between them. There were, as might be expected, some links that provided little if any insight into the stalled personal project—tropes are not always infallible as guides to action. But Poutine quickly became aware of several associations that seemed worthy of exploration. He suggested that his officers' lack of motivation might mean they were not receiving sufficient positive feedback when they had done something effectively. Most feedback is given in annual appraisals, and this would be rather like waiting until the end of the season before posting the results of the goals a hockey team scored (Senators, 417; Rangers, 287; Maple Leafs, 38; etc.). It doesn't quite have the motivational significance of an immediate red light and scoreboard replay. He also speculated that his best officers were seldom recognized for their team-work and that over time this could be demoralizing. He linked "friction with other officers" with the concept of an assist in hockey. An assist becomes part of the player's point total, carrying as much weight in the overall scoring statistics as a goal. Setting up other players' goals with-out getting any credit can be rather disillusioning.

Colonel Poutine took each of the ideas and reframed his "morale project" as one in which he should provide more frequent feedback on goals his officers achieved and should ask each officer in the annual ap-praisals to indicate other officers who had assisted them in their work. Although it would be an overstatement to say that creative metaphor analysis would become routine for Poutine, he did implement these two changes with positive results. Perhaps even more heartening was his realization that the same metaphor applied equally well to one of his other core personal projects: his son, who was negotiating the tricky terrain of living at home while saving for college. He too would benefit from more frequent feedback on his successes and greater acknowledg-ment of his value to the family. Although I don't know how that proj-ect worked out, I was touched that a simple exercise on work-related personal projects might have had positive implications for a father-son relationship.

WHO DO YOU THINK YOU ARE NOW?
PERSONAL CONSTRUCTS REVISITED

Another way of making our core projects more sustainable is by chang-ing the personal *constructs* through which we appraise them. Recall how, in Chapter 1, we discussed how our personal constructs provide both frames for the anticipation of events and cages within which we can become trapped. One of the creative therapeutic approaches that arose out of Kelly's personal construct theory is called Fixed Role Ther-apy (FRT).[6] To get a sense of what the therapy feels like to the client, imagine you have been seduced by your spouse into playing the role of the butler in your local community theater production of *Upton Abbey*, and the play's director is a passionate advocate of "method acting," which demands that you truly engage at the deepest level with the role. You learn to speak softly, to attend to detail, to be polite and discreet, to be solicitous in the extreme and ever alert to the need for unobtrusive intervention to make things seem effortlessly perfect. Then you begin to realize that this new role has started to smudge over into your real, off-stage, life. The new butler-ish you stands in marked contrast with your typical crash-bang-wallop approach to life, and as a result of enacting this role, you observe things you hadn't noticed before. And you also notice that other people react differently to *you*: they seem to listen more, to open up more, and to seek out your opinion on things other than NCAA basketball and artisanal beer. When the play ends and the role is abandoned, you find, at least for a while, that you are not entirely sure you want to go back to being the person you were.[7]

The same process occurs with FRT. It begins with the client writing a one- or two-page self-characterization, a version of which we described in Chapter 5. The personal sketch essentially answers the question "Who do you think you are?" On the basis of themes that emerge in this self-characterization, the therapist writes a sketch of a hypothetical person whom the client is to role-play over a two-week period. This script is, in a sense, a proposal about "who you might yet be." This script is carefully designed to invoke personal constructs that are "at right angles" to those the client typically employs; that is, they pull the client in new directions instead of rattling back and forth on constructs that no longer serve him well. The client and therapist review the kinds of situations and daily routines that will be met during the role playing and rehearse ways of acting and responding until the client is ready to move out on his own. By engaging in this role enactment, the client learns to view the world through a different set of goggles. The purpose of the FRT is not to create a permanent change in the client's personality—quite the opposite: it is to show him that he has the capacity to try on new possible selves that can free up paths of movement in his life.

Consider, for example, a man whose self-characterization is dominated by the personal construct of "idiot vs. genius." He applies this construct both to himself and others, and he deploys it in the all-or-none fashion that his choice of terms implies. Such a construct limits his degrees of freedom to pursue projects. He consigns himself to being a permanent idiot and sees no chance of moving over to the genius end of the spectrum, which he reserves for a very select few: his corporate lawyer mother, his superdorky younger brother, and Stephen Hawking. Of course this means that virtually all others with whom he comes in contact are also idiots, and his interactions with them reveal that he has such a stance. An appropriate script for him would be one in which he would be encouraged to construe others and himself in terms of the construct "skilled-unskilled." This more finely differentiated construct is likely to foster more possibilities in his pursuits: he and others can be skilled in some domains but not in others, and being skilled, unlike being a genius, is something that one can learn. By invoking this more viable construct he can notice pathways for change and personal development in both himself and others.[8]

SELF-PROJECTS: MATCHING, STRETCHING, AND SELF-DETERMINATION

In Chapters 2 and 3 we discussed how both relatively stable traits and free traits have important consequences for our well-being. Let's now see how they each relate to the sustainable pursuit of core projects.

First, people experience more positive project pursuit when there is a "fit" between their personality traits and their personal projects. Conscientious individuals, for example, tend to have meaningful and effective personal projects in a diversity of areas, including academic, health, and social domains, whereas neurotic individuals experience problems in those same areas. Extraverts, however, display a more specialized style—they are particularly happy and effective in projects involving interpersonal relations like hanging out with friends or going to exciting recreational events but are not as easily engaged when it comes to academic projects. Happiness is greatest in those for whom there is a convergence between their traits—the kind of personal projects they are pursuing and the themes they invoke when providing life stories. For example, we have found that individuals who have sociable traits are most happy if they are engaged in interpersonal projects and if their self-characterizations include themes of connection with others.[9] So understanding where you stand on some of the more stable traits of personality, besides being important in its own right, also helps you understand which projects you can pursue with the greatest likelihood of success and sustainability.

I have also made the case that we need to go beyond relatively stable traits to understand the course of our lives. We also need to understand *free traits*. We discussed how our core projects may require us to act out of character on occasion: a neurotic person may present himself as stable when fulfilling his professional functions, an introverted teacher may act as a pseudo-extravert when in front of her classes, or a highly agreeable community organizer may become strategically disagreeable, even fierce, to avenge a social injustice. Acting out of character through free traits fosters an increased likelihood of bringing a core project to fruition. Free traits make us stretch and grow. For example, recent research suggests that asking introverts to act in an extraverted fashion actually increases their positive mood and well-being. This is an interesting example of the benefits of acting out of character.[10] But I speculate that if this kind of behavior is engaged in over a protracted period of time, it may well take a toll.

Protractedly acting out of character then may not be sustainable. So what might we do to mitigate the costs of acting in this fashion? We talked earlier about the value of finding restorative niches in which we can regain our first natures and indulge our biogenic selves. But what about those cases in which acting out of character isn't just an occasional thing or a short-term adaptation to situational demands? What happens when we decide to actually change our traits, when our projects are self-change ventures in which we commit ourselves to "be less assertive" or "more extraverted" or "stop being such a jerk"?

These intrapersonal projects or, more simply, self-projects have important and seemingly paradoxical effects on the quality of our lives. Independent studies carried out in Finland and North America show that individuals who were engaged in self-projects, relative to other kinds of projects, tended to experience more feelings of depression.[11] Why would people concerned about "improving" themselves be given to depression? One reason is a tendency for such projects to become ruminative concerns, but another reason is that self-projects are also typically rated as low on efficacy—we doubt they will be successfully completed. Because a belief that our projects will succeed is a key aspect of well-being, we might be tempted to encourage ourselves and others to "get over yourself" and focus on more promising pursuits. But we should be cautious with such admonishments. Self-projects also have interesting links with creativity. More creative individuals are likely to identify with their self-projects and construe them as being exploratory ventures rather than depressing burdens.[12]

Why do some individuals struggle with the self and others find self-exploration to be invigorating? One possible answer is the origin of the self-project. Consider two individuals with the same personal project of "being more outgoing." For one, the origin of the project was external—it came from her boss in the sales department who insisted that a change was needed because sales were down and she scarcely had an animating presence when meeting customers—indeed, she "had absence," he told her cruelly. She could either go find a job with a better fit to her withdrawn personality, or she could change—her choice. But consider another person, someone with the same project of being more outgoing. In her case it arose out of her own reflection on how many of the things that mattered most to her required getting over an initial hesitancy in engaging with others and required her to push herself to be more outgoing. She tries some small experiments to see whether she could do this on a trial basis, and it worked. She sees this as a rather intriguing process of stretching herself, and it feels good. She now sees it as a core project—her choice.

In the first case the choice was externally imposed and accompanied by an implicit " . . . or else." In the second it originated from within the person herself. It was an internally generated self-expression. There is a very strong reason to believe that the internally generated project will fare better than an externally generated one. The explanation flows from self-determination theory, a highly influential theory of personality and motivation that contrasts internal, autonomously regulated goals with externally regulated ones.[13] The internal self-generated projects are more sustainable and provide greater benefits for emotional and physical well-being than do those that are external and controlling. So with respect to the seemingly paradoxical findings that self-projects are both linked to depressive feelings and to creativity, it is plausible to suggest that their origin is critical. When changing or challenging the self is regarded as a personal initiative rather than an external imposition, it is likely to be more meaningful, manageable, and sustainable.

CONTEXT MONITORING: SCANNING, SEEKING, AND SHAPING OUR ENVIRONMENTS

These initial strategies for sustaining core project pursuit involve ways of reframing or creatively reconstruing our projects and of mounting self-change projects that can advance our well-being. In essence, they focus exclusively on the individual person—on you, yourself. But focusing only on your singular self and ignoring the environments within which you engage with life is unduly restrictive. So the middle chapters of the book focused our attention on the environmental contexts of our lives: the situations, places, cities, and social ecologies in which we pursue our core projects.

We looked at the subtleties of having a sense of control or agency in our lives and how it contributes to our emotional and physical well-being. We saw that a sense of control is generally a positive thing. But we concluded that it was adaptive *only* if based on an accurate reading of the actual environmental contingencies within which our lives are embedded. Are those buttons hooked up? Are those aspirations based on an optimal degree of illusion? In short, have we scanned our context properly?

Perhaps you have a child who is away at college for the first time. You have a core project of weighing in with advice, providing lots of love and support, and, oh yeah, cash. Again and again. But being able to sustain that core project requires you to scan the ecosystem carefully. Now that it is deep in December, is your child the same one who headed off in September? Is he looking different—really different? Does he have new friends? Do they come with benefits? Or costs? Are you still insisting he takes vocationally relevant courses even though the economy is improving and he has fallen in love with medieval history? If we do not scan for updates, we run the risk of engaging in projects that may fit well with our initial aspirations but are unsustainable because the social ecology has changed. In short, accurate scanning increases the vigor and viability of project pursuit.

Contexts are not merely constraining; they can also potentiate the pursuit of what matters to us. We saw that some situations generate the scripts through which we enact our goals and desires, and some individuals, the high self-monitors, are particularly sensitive to engaging with those situations. And we observed the same function being fulfilled on a larger scale by the kinds of cities and regions to which we are attracted. Here again we saw how individuals actively seek out a good fit between their personalities, their core projects, and their environments.

The concept of a niche is helpful in understanding the relation between people, projects, and places. We first discussed niches in the context of the need for individuals to find solace from acting out of character and the value of finding restorative niches in which they can reclaim their biogenic natures. But a restorative niche is just a special adaptation of a more generic kind of niche—let's call it an *identity niche*—in which we find an optimal fit between our interests, traits, aspirations, and places that afford them expression. When stimulus-seeking extraverts choose to move to an exciting metropolis or an anxious introvert finds a special place in the library that is safe and soundproof, we are witnessing niche-seeking in action.

But niches, at least in the sense that ecologists use the term, have another feature that is relevant here. Niches are contestable. They are typically occupied by members of one species and defended against occupation by other species, but there is often also competition within the species for access to the niche. The same kind of dynamic occurs in human families too. Perhaps you have puzzled over how very different two siblings are from each other and had the passing pernicious thought that although you and your sister have been told you had the same mother (and they showed you pictures), you really suspect you had different fathers. In his book *Born to Rebel*, Frank Sulloway developed the argument that family dynamics are contests in which children compete for parental resources by occupying niches and defending them.[14] Under this theory first-born children have the initial pick of the niche, whereas later-born ones need to create and find their own. First-born children, according to Sulloway, are characterized by conservative traits like conscientiousness and neuroticism, and this means they are likely to be rule-following and to adopt their parents' values. Later-born children face a dilemma. They need to compete for parental attention and resources with a sibling who is bigger and stronger and who, in some respects, can play a quasi-parental role to the younger sibling. Given the difficulties of competing directly for the niche the older sibling already occupies, the later-born children adopt another strategy: they create their own niches. Instead of being the conscientious, careful, traditional one, they become the exploratory, norm-bucking, and potentially more rebellious sibling.

But if this theory of family niche dynamics is right, it poses an interesting question: What if the later-born has a biogenic tendency to be careful and cautious and compliant? If that niche is taken in the family by an older sibling, then the later-borns, in order to create a new niche, may need to act out of character. Their niche strategy may require the long-term enactment of free traits rather than the natural expression of inborn stable dispositions. As a consequence, later-born individuals will need to find restorative niches in order to mitigate the cost of acting out of character more than their older siblings do. There may well be a very sound adaptive reason why that younger brother of yours so fiercely resisted your attempts to take over his secret hideaway or bristled when you questioned whether he really was the family rebel.

SELF-REFLECTIONS: RECONCILING AND REVITALIZING

So here we are, wrapping up. Here's where we have gone. We talked about the personal constructs that can frame our experience but create cages out of which we need to escape. We explored the links between our relatively stable traits and important life outcomes and also how we can engage in free traits in order to advance what matters most to us in our lives. We saw how a sense of agency in our lives has positive consequences but only if accompanied by alertness to the realities of our environments. We cautioned that an overzealous lifestyle could be dangerous to our health unless mitigated by a sense of play. We examined how the situational demands of our lives shape self-presentation for some people but not for others. We observed how creativity requires audacious imagination and commitment but also an awareness of others' unacknowledged contributions. We surveyed how geography and personality are intertwined and how certain personalities are drawn to certain cities and regions. We considered the importance of how personal projects, especially core projects, provide our lives with a sense of meaning and structure, connection and emotional richness. And we saw how such projects can lose meaning, go stale, and then be revitalized.

It remains for us to stand back a bit and ask you how you are doing. Have you been reflecting on your own life as you have gone through the chapters? You have likely had and will in the future have other occasions on which to reflect on your personality and how your life has gone. Such moments of reflection often occur at major transition points in our lives. A graduation, a marriage, a divorce, a promotion, a job loss, a retirement—all of these call for some reflection on how we are doing, where we are going, and how to proceed. They are linked to different stages of the life cycle and are expectable, if not always desired, occurrences in modern life. Other calls for reflection are subtler. You have a quiet conversation with a friend who asks how you are *really* doing, and you are taken aback that you hesitated so long before answering. A friend dies, you are asked to give a eulogy, and it brings you to tears, but not only because you are grieving: you kept seeing yourself in the sentences you were reading about him. You have had a sleepless night cursing that infuriating dragon who messed up the meeting this morning—just who did she think she was?—and then conceding that she was, in fact, you. How can you reconcile these two different selves at two in the morning?[15]

In these reflections you may have seen that you and yourself play somewhat different roles. You construe yourself through personal constructs, and the self you construe might be closed and afraid to venture out. Or you act out of character and leave your comfortable self behind. Or you find yourself in a situation that has such a powerful and unwanted influence on you that you really weren't yourself at all. Perhaps you have lost yourself in a core project that made you transform your sense of who you truly were, creating a new self in the process. Each of these transactions between you and yourself can be illuminating, but they can also be challenging—you and your self may need some reconciliation.[16]

Owen Flanagan, a distinguished philosopher who has thought deeply about personality and well-being, has written an account of the theme of self-reconciliation. It appears in a scholarly book called *Self Expressions* and is a two-page epilogue. His metaphor for such moments of reconciliation is of a final dance between an "I" and the self it observes: between you and yourself or me and myself. He ends with an invocation to his self. If you read it slowly, out loud, it could be mistaken for poetry.

Darling self, it is inevitable, but it is also my wish, that you
save the last dance for me. Never mind the clumsiness, we know
each other well by now. And let's hope—this seems both

romantic and reasonable—that value is detected as we embrace . . .
But it should be more than mere infatuation. It should
really matter. It, this life, that is, should really mean something.
Respect, even self-referential respect, should be warranted. It will be
good to feel peace of mind, to be comfortable, to sense integrity, and
effort, and to recognize that we have had some fun. Remember, if anyone
knows you—really remembers and knows you—especially how you dance,
it is me. Me, myself and I. Cha, cha, cha.[17]

This is, of course, a poignant treatment of the very theme with which we began this final chapter—reflecting on life, getting things in order while there is still time to do so. Don't think of this as a sad or distressing image—it is quite the opposite. Our dances with selves can take place at any time; we needn't wait until we are old and wobbly to reconcile ourselves to ourselves. And although self-reflection may begin with reconciliation, it also creates the impetus for revitalizing our lives. With some new personal constructs for thinking about yourself and others as well as a deepened self-awareness, you may have already been dancing through these chapters and thinking, "That's me, that's my very self."

Flanagan's dance is a pas de deux between the internal you and the enacted self that you have constructed and nourished and occasionally fought with throughout your life. It is both a reconciliation and an evocative account of how we might look back on a life well lived. Read it again, slowly. I find the juxtaposition of "integrity and effort" with having some fun deeply appealing. For our lives to be meaningful we need to commit to core projects and pursue them passionately. But such pursuits need to be counterbalanced with a touch of lightness and whimsy, or else the whole venture can flounder.

We can stretch the dance metaphor further to include two other constituencies of the self. First, for many of us there is more than one self to consider. You high self-monitors now know who you are. And many of your selves have never been introduced. Can you choreograph a dance between your "professional woman" self and the loopy you who takes too many selfies while eating cold pizza in bed on a Sunday morning? Can these alternative selves at least hold hands, if not do a tango together? Or maybe you are a guy whose favorite self is a "guts and glory" man, but you worry that your timid meerkat will pop up its head and ruin your well-crafted image. Maybe you can find a way of merging manliness with vulnerability?

And the second invitees to our dance should be other people who have mattered to us along the way, the ones who shape our expectations, support our ventures, and love us in spite of ourselves. So here's to you, yourself, by all means, but also here's to *us*, your fellow travelers in life, who help shape your personality, promote your well-being, laugh at your jokes, and hold you tight when you most need it.

Acknowledgments

I STARTED THIS BOOK IN 2000, WHEN I WAS IN THE INAUGURAL class of the Radcliffe Institute for Advanced Study at Harvard University. I am very grateful for the support and stimulation provided by this remarkable institution. *Me, Myself, and Us* is based on lectures I have developed over many years, and I wish to thank my students at Carleton, Harvard, McGill, and Cambridge Universities for encouraging me to write it and thank my colleagues for their support. In particular I would like to acknowledge the stimulation of students in my Social Ecology Lab at Carleton, the HAPPI group at Harvard, and the Social Ecology Research Group (SERG) at Cambridge. For many years Carleton supported my research and provided a cordial collegial environment. Ai-Li Chin, Adam Grant, and Susan Cain have each had a vital role to play in shaping my ideas and aspirations for the book, and they are de facto lifetime honorary members of our Social Ecology Research Group. They are also splendid friends.

Recently I have been the beneficiary of a generous visiting fellowship at Sidney Sussex College, Cambridge, and I wish to thank Michael Lamb, who made it happen and for his inspirational leadership in the Social and Developmental Psychology group at Cambridge. Professor Felicia Huppert has been an enthusiastic advocate for the ideas in this book, and I am most grateful for her collegiality and friendship. I am fortunate to be a fellow at the Well-being Institute at Cambridge, founded and directed by Felicia, and I look forward to continuing our explorations of how the science of personality and the art of well-being can gain from each other's perspective.

My editor at PublicAffairs, Lisa Kaufman took on the onerous task of herding my different aspirations in writing this book and provided thoughtful and sensitive counsel at every stage of the venture. Thanks, Lisa, for helping me keep the reader foremost in my thoughts and tempering my strong tendency to have my dear students keep popping up and dominating the discourse. Harvey Klinger, my agent, was especially helpful, and on more than one occasion he saved me from myself. Thanks, Harvey, for your expert advice and good cheer. During the early stages of writing, Deanna Whelan was my spirited and meticulous research assistant. During the wrapping-up stage of the project I was very fortunate to have Simon Coulombe provide research assistance. He is the perfect blend of conscientiousness and generosity. Merci bien, Simon. My thanks also to the remarkable production and marketing folks at PublicAffairs and Perseus Books Group, who showed patience and skill in abundance.

And so to a particular pleasure—thanking my family. My late parents, Ada and Richard Little, deeply wise and wonderfully warm folks, would have loved knowing that I actually finished the book, having seen me flailing away at it in various ways since I was six years old—or so it must have seemed. My sister, Margaret, from the outset, literally, saw the promise in what her little brother was doing, for which I will always be grateful. My children, Hilary and Benjamin, in their own splendidly singular ways have enriched my life and this book, providing advice, encouragement, and anecdotes galore. Thanks, too, to my extended family, the Littles, Parkers, Little-Hens, and their spouses and children, particularly Steve, Clover, Finn, and Susan. They have been deeply supportive and enthusiastic as I have worked on this project, and I thank them most deeply.

My wife, Susan Phillips, has been the single-greatest source of inspiration for *Me, Myself, and Us*. She was unstinting in encouraging me when things weren't going well and rejoicing with me when things were sailing along. She helped me see the logic in my own ideas, sharpened my thinking, nurtured my confidence, and showed me how to hang in when I was about to chew off my elbows. She is brilliant and golden and the love of my life.

Notes

NOTES TO PREFACE

1. I am a big believer in footnotes. In this I differ from the inimitable Professor Daniel Gilbert (2006), the author of *Stumbling on Happiness*, who, in his first footnote, suggested that it was the only important one to read of the several hundred following it. I want you to read most of mine because they allow me to add nuance, subtlety, and shading to the bolder assertions of the main text. This is not to say, whatsoever, that Professor Gilbert isn't nuanced, subtle, or shady. After reading Chapter 2 you will have a better insight into why I prefer to write—and perhaps you prefer to read—the small print. We differ in predictable ways from those of you who are currently not reading this. Now for the reference you were seeking when you turned here:

For a recent authoritative treatment of the burgeoning field of positive psychology see Seligman (2011). I have written elsewhere on the relation between personality science and positive psychology in Little (2011).

NOTES TO CHAPTER 1

1. The phrase "frames and cages" as applied to personal constructs is taken from Ryle (1975).

2. See Kelley and Michela (1980) for an early review of the attribution literature.

3. For this section on familiar strangers and frozen relationships, I have drawn heavily on Stanley Milgram's (1970) original account.

4. Dan McAdams (1995) has presented a compelling case for viewing personality as a three-tiered structure with traits, personal concerns (or projects),

and narratives representing the ascending tiers. He presents a fascinating and highly readable illustration of these levels in his analysis of George W. Bush (McAdams, 2010).

5. Kelly's (1955) personal construct theory was an audacious and highly innovative approach to the study of personality. It anticipated by at least a decade the cognitive turn in psychology, and it remains influential in personality psychology, clinical psychology, and organizational studies. For comprehensive reviews of personal construct theory and its applications, see Fransella (2003) and Walker and Winter (2007). As an undergraduate I came across Kelly's book when searching out a reference work on neuropsychology. Instead of finding the *Stereotaxic Atlas of the Brain*, a badly shelved copy of *The Psychology of Personal Constructs* appeared in its place. I started to look through it, sat down on the library floor, and, four hours later, emerged a Kellian and shifted my doctoral studies from neuropsychology into personality psychology. As we'll discuss later, such chance encounters play a significant role in the course of our lives (Little, 2007).

6. For a recent helpful account of the role of emotions in personal construct theory, see Lester (2009).

7. Hostility is to be distinguished from aggression, which, for Kelly (1955), is simply the active elaboration of your construct system. In this respect aggression is not seen as a negative aspect of personality so much as a creative, active, exploratory stance toward events in your life.

8. The original work on core constructs, implicative richness, and resistance to change was presented by Dennis Hinkle (1965) in what has become a classic in the personal construct literature.

9. Throughout the book, whenever discussing individuals or organizations, I have changed the names and altered some of the details and circumstances so anonymity is maintained.

10. For details on person-thing orientation and specialization theory, see Little (1972, 1976).

11. See Little (2005) for a discussion of how these contrasting approaches play out in the field of personality science.

12. Assessment centers are not fixed places but events that organizations put together that are typically held away from the workplace. The ratio of the candidates to the assessors is typically very high, often 1:1, and the assessors are divided evenly into those who have specialist training in personality and abilities assessment and those who are employees of the organization and have extensive experience with the type of position being evaluated. Prior to the

event candidates are administered a battery of personality, ability, and interest tests. For a recent comprehensive review on the functions and validity of assessment center methods, see Duncan, Jackson, Lance, and Hoffman (2012).

13. We deal extensively with Personal Projects Analysis in Chapters 9 and 10.

14. Independent of our own research, Robin Vallacher and Dan Wegner were also examining the same issues in what they called action identification theory (Vallacher & Wegner, 1987).

15. See Little (2005).

16. For those who are interested in delving deeply into an analysis of their own personal constructs, there are assessment techniques available for that purpose. An excellent resource on repertory grids, the technique for assessing personal constructs, is found on the website run by the University of Hertfordshire in the UK, www.centrepcp.co.uk.

NOTES TO CHAPTER 2

1. I use the convention adopted by personality psychologists of spelling it extraversion, rather than as extroversion, which is preferred by dictatorial spell-check programs.

2. The ancient origins extend back to the pre-Socratic Greeks (see Dumont, 2010; Winter & Barenbaum, 1999).

3. Of Jung's contributions, it was his treatise on psychological types that had the greatest influence on the MBTI. See Jung (1921).

4. For details on the MBTI, see Myers, McCaulley, Quenk, and Hammer (1998).

5. See Pittenger (1993) for an overview of some of the issues regarding the reliability and validity of MBTI profiles.

6. A highly critical account of these issues is found in Paul (2004).

7. Cited in Zemke (1992).

8. Karl Scheibe (2010) has written a fascinating account of how MBTI workshops engage the participants in an act of high drama, akin to the kinds of performances offered by magicians.

9. An excellent and accessible introduction to the Big Five dimensions of personality is found in Nettle (2007). I have inferred my own biogenic tendency toward introversion on the basis of an early model of Eysenck (1967) based on differences in levels of neocortical arousal in introverts and extraverts. I should note, however, that more recent research places greater emphasis on the effects of neurotransmitter activity on introversion-extraversion. See DeYoung (2010).

10. See the original article by Gosling, Rentfrow, and Swann (2003). How to score the TIPI:

Conscientiousness:
    Score for #3: _7_ /
    + (8 – Score for #8): _6_ |7
    = _13_
    Divide your answer by 2.
Conscientiousness = _6.5_ /7

Agreeableness:
    Score for #7: _6_
    + (8 – Score for #2): _6_ |4
    = _12_ |10
    Divide your answer by 2.
Agreeableness = _6_ /5

Emotional Stability: *(Note: low scores are associated with Neuroticism)*
    Score for #9: _7_ |7
    + (8 – Score for #4): _1_
    = _14_
    Divide your answer by 2.
✓ Emotional Stability = _7_

Openness to Experience:
    Score for #5: _2_ |4
    + (8 – Score for #10): _2_ |3
    = _4_ |7
    Divide your answer by 2.
Openness to Experience = _2_ |3.5

Extraversion:
    Score for #1: _2_ |5
    + (8 – Score for #6): _3_ |2
    = _5_ |7
    Divide your answer by 2.
Extraversion = _2.5_ |3.5

Adult Average Scores: *(Based on 305,830 participants. My thanks to Jason Rentfrow for providing this information.)*

	Average	
Conscientiousness	4.61	High Scores = 6.0 and above *6.5*
		Low Scores = 3.2 and below
Agreeableness	4.69	High Scores = 5.9 and above *6*
		Low Scores = 3.5 and below
Emotional Stability	4.34	High Scores = 5.8 and above *7*
		Low Scores = 2.9 and below
Openness to Experience	5.51	High Scores = 6.6 and above *2*
		Low Scores = 4.4 and below
Extraversion	3.98	High Scores = 5.6 and above *2.5*
		Low Scores = 2.4 and below

Gosling, S. D., Rentfrow, P. J., & Swann Jr., W. B. (2003). A very brief measure of the Big-Five personality domains. *Journal of Research in Personality*, 37(6), 504–528. Elsevier Science. Reprinted with permission.

11. See Costa and McCrae (1992) for the latest edition of the NEO PI-R and a shorter version used frequently in personality research. It should be noted that the NEO PI-R includes six facets for each of the major traits, whereas the shorter version measures only the Big Five traits. Lewis Goldberg at the Oregon Research Institute has developed a remarkable public domain resource known as IPIP: The International Personality Item Pool (ipip.ori.org) provides a large number of personality scales and should be consulted by those interested in the more technical aspects of personality scales. For those specifically interested in assessing their own Big Five scores, based on IPIP, see John A. Johnson's very helpful site: http://www.personal.psu.edu/j5j/IPIP/ipipneo120.htm.

12. There are several excellent summaries of the rise and influence of the five factor theory of personality traits. For a recent, authoritative source see John, Naumann, and Soto (2008).

13. In a study of identical and fraternal twins measured with the NEO PI-R the genetic influence was estimated to be: neuroticism (41%), extraversion (53%), openness (61%), agreeableness (41%), and conscientiousness (44%) (Jang, Livesley, & Vernon, 1996).

14. The definitive review of research on the consequential impact of personality traits on different aspects of well-being, including achievement, health, and happiness, is Ozer and Benet-Martínez (2006).

15. See, for example, Bogg and Roberts (2004) and Barrick and Mount (1991).

16. See McGregor, McAdams, and Little (2006) for a discussion of the different trajectories of conscientious vs. party-animal students.

17. See Friedman et al. (1993).

18. See Nettle (2007).

19. See Hogan and Hogan (1993).

20. I draw here on an extremely informative book on the interactions of jazz musicians and house bands who frequently, without knowing each other, need to craft performances that sound seamless (Faulkner & Becker, 2009).

21. See David Buss's definitive text on evolutionary psychology (2008). For an authoritative review of the research on agreeableness more generally see Graziano and Tobin (2008).

22. See Judge, Livingston, and Hurst (2012).

23. See Mahlamaki (2010).

24. See Barefoot and Boyle (2009).

25. See Booth-Kewley and Vickers (1994).

26. See Moskowitz and Coté (1995).

27. See Steel, Schmidt, and Shultz (2008).

28. See Widiger (2009) for a detailed analysis of recent research on neuroticism.

29. There is evidence in a major longitudinal study for the long-term impact of neuroticism on physical health. See Charles, Gatz, Kato, and Pedersen (2008).

30. Also, Samuels and Widiger (2011) have presented evidence that obsessive-compulsive personality disorder is an extreme form of the "normal" trait of conscientiousness. I suspect that neuroticism will accentuate this progression.

31. See, for example, Buss (1991) and Nettle (2006) on evolution and the adaptive significance of the full range of personality dimensions.

32. For a detailed analysis of openness to experience, see McCrae and Sutin (2009).

33. McCrae (2007) first reported the link between pilo-erections and openness to experience.

34. Cain sees introverts as occupying the same subservient role in American society as women did at the beginning of the modern women's movement, and the popularity of her spirited portrayal of the strengths of introverts has created a "quiet revolution." See Cain (2012).

35. The neocortical arousal model of extraversion was initially advanced by Eysenck (1967). A recent authoritative review of research on extraversion can be found in Wilt and Revelle (2009).

36. See Loo (1979).

37. Lynn and Eysenck (1961), but see Barnes (1975) for more equivocal results.

38. See Revelle, Humphreys, Simon, and Gilliland (1980) and Wilt and Revelle (2009).

39. See Grant (2013).

40. Much of the following section draws on the comprehensive review of extraversion by Wilson (1978).

41. For a recent and valuable account of the links between caffeine, extraversion, and memory see Smith (2013).

42. See McAdams (2009).

Notes to Chapter 3

1. See Gosling (2009).

2. On counter-dispositional behavior, see Zelenski, Santoro, and Whelan (2012).

3. An excellent review of this literature is found in Roberts and DelVecchio (2000).

4. In this chapter I draw extensively on a chapter I wrote with Maryann Joseph (Little & Joseph, 2007).

5. See, for example, DeYoung (2010).

6. See, for example, Elliott (1971). There is also evidence that extraverted parents' newborns also seek out more auditory stimulation compared to those of introverted parents (Bagg & Crookes, 1975).

7. The technique was first reported in Eysenck and Eysenck (1967). Regarding the need to use concentrated lemon juice, see Howarth and Skinner (1969).

8. See, for example, von Knorring, von Knorring, Mornstad, and Nordlund (1987).

9. For a review and research agenda on the biological basis of traits, see DeYoung (2010).

10. See Kogan et al. (2011).

11. Although there is widespread consensus that American and Western European countries differ on the introversion-extraversion dimension from countries such as Japan, the psychological evidence for this is not extensive. Some confirmation of the bias toward extraversion in the West, however, is found in Schmitt et al. (2007).

12. See Triandis and Suh (2002).

13. See Little (1996) and Little and Joseph (2007).

14. See my account of this in Lambert (2003).

15. See McGregor, McAdams, and Little (2006).

16. See Roberts and Robins (2003).

17. See Cantor, Norem, Niedenthal, Langston, and Brower (1987).

18. The pioneering work on emotional labor was done with airline attendants by the sociologist Arlie Hochschild (1983).

19. This section draws heavily on results reported in Jamie Pennebaker's intriguing research reviewed in his engaging book *Opening Up* (1990).

20. See Pennebaker, Kiecolt-Glaser, and Glaser (1988).

21. The late Dan Wegner was one of psychology's most gifted and creative researchers. See his wonderful book *White Bears and Other Unwanted Thoughts* (1989).

NOTES TO CHAPTER 4

1. Snyder, M., & Gangestad, S. (1986). On the nature of self-monitoring: Matters of assessment, matters of validity, *Journal of Personality and Social Psychology*, 51(1), p. 137. American Psychological Association, Washington, D.C. Reprinted with permission.

2. I am grateful to Mark Snyder for providing me with updated norms for adults.

3. See Snyder (1974, 1979).

4. I first came across this in a quirky video by Richard Wiseman ("Are You a Good Liar? Find Out in 5 Seconds," YouTube, www.youtube.com/watch?v = yRAmvLV_EmY&list = PLy9A-KHMzTjh9CY4JafXD7fsJey25Awzd), who used the letter Q. However, he cites an earlier article by Hass (1984) that used the letter E to make the same point.

5. See Mischel (1968).

6. Mischel was a student of George Kelly's, whose personal construct theory we discussed in Chapter 1. Although the impact of Mischel's treatise was originally seen as emphasizing situational factors over personality factors, he was equally concerned with showing that the way we construe our lives was critical, a clear indication of the influence of Kelly.

7. For reviews of the person-situation debate that detail some of the complexities that arise when we try to estimate the relative importance of personality and situational factors, see Argyle and Little (1972), Endler and Magnusson (1976), and Little (1999a).

8. See Snyder (1974).

9. These studies and much that follows in this section are detailed in Snyder's influential and highly readable book *Public Appearance, Private Realities: The Psychology of Self-Monitoring* (1987).

10. See Snyder and Gangestad (1986).

11. See Snyder, Gangestad, and Simpson (1983).

12. See Snyder and Simpson (1984, 1987). For a comprehensive review of studies examining self-monitoring and personal relationships, see Leone and Hawkins (2006).

13. See Kilduff and Day (1994).

14. Ibid.

15. Ibid.

16. This research is reported in a creative and groundbreaking article by Wallace (1966). See also the research by Paulhus and Martin (1987).

17. See Turner (1980).

18. This is introduced in Murray's classic *Explorations in Personality* (1938).

19. I thank Max Gwynn and Hans de Groot, who provided creative input to this study.

20. See Snyder (1979).

21. The term was coined by the distinguished psychiatrist, Vivian Rakoff, in a political context but I believe his concept has much broader applicability, particularly with respect to self-monitoring theory.

NOTES TO CHAPTER 5

1. This is the Personal Control scale taken from Del Paulhus's Spheres of Control instrument (Paulhus, 1983), and I thank him for his permission to use it for illustrative purposes. Paulhus, D. L. (1983). Sphere-specific measures of perceived control. *Journal of Personality and Social Psychology*, 44(6), 1253–1265. American Psychological Association, Washington, D.C. Reprinted with permission. doi: 10.1037/0022-3514.44.6.1253.

2. The classic paper on locus of control was written by Rotter (1966). In this chapter I have drawn on comprehensive reviews by Phares (1965) and Lefcourt (1982). For an excellent recent review, see Furnham (2009).

3. See Asch (1940).

4. See Crowne and Liverant (1963).

5. This revealing study was reported in Biondo and MacDonald (1971).

6. See Platt (1969).

7. See Lefcourt (1982).

8. See MacDonald (1970). Note this research was carried out when there was much more variability in birth control practice than at present.

9. See Coleman et al. (1966).

10. For example, see Ng, Sorensen, and Eby (2006).

11. See Seeman (1963).

12. See Mischel, Ebbesen, and Zeiss (1972).

13. See Casey et al. (2011).

14. These classic studies were carried out by Glass and Singer (1972).

15. See Langer and Rodin (1976).

16. See Schulz and Hanusa (1978).

17. Julie Norem has done some important research on the differences between illusory glow optimism and defensive pessimism. See Norem (2002).

18. The definitive study of the positive effect of illusions is in Taylor and Brown (1988).

19. This section draws on the important work of Peter Gollwitzer and his colleagues (e.g., Gollwitzer and Kinney, (1989). They have examined the impact of deliberative vs. implemental "mind-sets" on illusion, a distinction that I believe is similar to the initiation and implementation stages of project pursuit.

NOTES TO CHAPTER 6

1. This table is taken from "The Social Readjustment Rating Scale," Thomas H. Holmes and Richard H. Rahe, *Journal of Psychosomatic Research*, Volume 11, Issue 2, August 1967, pages 213–218. Copyright © 1967. Published by Elsevier Science, Inc. All rights reserved. Reprinted with permission from Elsevier.

2. See Vinokur and Selzer (1975).

3. See Maddi and Kobasa (1984).

4. Thanks to Michael Scheier for permission to reproduce the photo which appeared in Carver and Scheier (1992).

5. An authoritative and comprehensive account of research on hostility and Type A behavior has been written by Barefoot and Boyle (2009).

6. See Wegner (1994).

7. See Antonovsky (1979).

8. See Caspi and Moffitt (1993).

NOTES TO CHAPTER 7

1. Apparently Lady Gaga phoned this message on creative music production from Dublin to *Blender* magazine. See Sarah Zashin-Jacobson, "Lady Gaga in Blender," Examiner, March 10, 2009, www.examiner.com/article/lady-gaga -blender.

2. Gough, H. G. (1979). A creative personality scale for the Adjective Check List. *Journal of Personality and Social Psychology*, 37(8), 1398–1405. American Psychogical Association, Washington, D.C. Reprinted with permission.

3. There are several key publications summarizing the results of the IPAR studies. In this chapter I draw heavily on the results presented in MacKinnon (1962) and Barron (1963).

4. This section draws on research exploring the relation between narcissism and creativity in Goncalo, Flynn, and Kim (2010).

5. This essay, by Kenneth Rexroth, originally appeared in *The Nation* and can be accessed at "My head gets tooken apart," *The Nation*, December 14, 1957, www.bopsecrets.org/rexroth/essays/psychology.htm.

6. The current version of the SVIB is called the Strong Interest Inventory and is an excellent resource for those who wish to see how their vocational interests compare with those who have succeeded in different fields. It is particularly helpful for those exploring possible occupations or a change of jobs. The online version also has a useful guide on how to interpret your profile and incorporate the results into your exploration of vocational possibilities in your life. Access it through Your Life's Path, Strong Interest Inventory, www .personalityreports.com/?view = Assessments_strong&gclid = CJ7QkOWS07 sCFawRMwodhzQABw.

7. IPAR did study women in some of their other studies of creativity. A particularly notable set of studies on women mathematicians was carried out by Ravenna Helson. See, for example, Helson (1971).

8. The study of psychological androgyny, in which separate scales of masculinity and femininity were used, was pioneered by Bem (1974). For evidence of the relation between creativity and androgyny, see Jönsson and Carlsson (2000).

9. Remember my earlier allusion to the poet Kenneth Rexroth's critique of his experiences as a participant in the IPAR creativity studies? Here is what he had to say about the creators of the MBTI :

"It was probably the best autoanalysis of two Jungian ladies ever done.

I wouldn't like to know them. They were mortifyingly shy in mixed company.

They said the wrong things when out socially and then regretted them bitterly in the wee hours. They didn't like their complexions. They didn't like men. Not even Carl Jung. They were real foul-ups. You could tell from the questionnaire. Fortunately they didn't appear in the flesh, just their most distressing questionnaires."

Rexroth displays an appalling ignorance of psychological assessment as well as a sexism that was typical of his day. But he does convey a view with which we will need to contend. Some people have an intense skepticism about certain types of psychological tests, and highly creative people are likely to be among them.

10. In Chapter 3 we dealt in depth with the issue of how social engagement may be depleting for some people. For an example of how this applies to a highly creative person, see Little (2007).

11. See Weeks and James (1995).

12. The MMPI Ego-Strength scale was developed by Barron (1953) to differentiate between those who benefitted and those who did not benefit from psychotherapy. High scores are found with individuals who have active coping styles and adequate social skills, among other positive attributes. Those scoring low on ego-strength will have difficulty making it on their own when under stress.

13. See Peterson, Smith, and Carson (2002).

14. See Carson, Peterson, and Higgins (2003).

15. See, for example, MacKinnon (1965).

16. For some compelling research on the intrinsic motives underlying creativity and the possible costs of adding external inducements, see Hennessey and Amabile (1998).

17. For Pickering's account of Darwin's illness, see Pickering (1974).

18. This appeared in a letter Darwin wrote (March 31, 1843) to Captain Robert Fitzroy, the commander of the *HMS Beagle*, several years after Darwin had returned from the voyage. The citation appears in Pickering (1974, p. 74).

19. From Tim Berra's authoritative account of Darwin's family life (Berra, 2013, p. 22).

Notes to Chapter 8

1. See Alexander (1964).

2. See "Contrasting Concepts of Harmony in Architecture: The 1982 Debate Between Christopher Alexander and Peter Eisenman," Katarxis № 3, www .katarxis3.com/Alexander_Eisenman_Debate.htm.

3. See Alexander (1970).

4. The original article is Milgram (1970). I have written about the contrast between these two conceptions of urban space elsewhere (Little, 2010).

5. See, for example, research on the "Type A city" by Levine, Lynch, Miyake, and Lucia (1989).

6. An informative account of the personal background to Milgram's subway studies appears in Michael Luo, "'Excuse Me. May I Have Your Seat?'" New York Times, September 14, 2004, www.nytimes.com/2004/09/14/nyregion /14subway.html?pagewanted = all&_r = 0.

7. I have written a detailed analysis of the relation between personality psychology and environmental psychology that expands on a number of the themes in this chapter (Little, 1987b).

8. McKechnie, G. E. (1977). The Environmental Response Inventory in application. *Environment and Behavior*, 9(2), 255–276. Reprinted with permission.

9. See Florida (2008). Florida's Who is your city? website contains additional material of interest to those wishing to examine their environmental preferences, www.creativeclass.com/_v3/whos_your_city.

10. See Rentfrow, Gosling, and Potter (2008).

11. For a thoughtful report on the consequences of the oil bonanza in North Dakota, see Rosanne Kropman, "How Oil Fracking Transformed a Poverty-Hit Prairie Town," Telegraph, February 21, 2014, www.telegraph.co.uk/earth /environment/10651934/How-oil-fracking-transformed-a-poverty-hit-prairie -town.html. It should also be noted that the population surge in towns like Williston, North Dakota, is overwhelmingly male.

12. See the greetings on their community website: "Always Warm!" Fargo-Moorhead Convention and Visitors Bureau, www.fargomoorhead.org/index.php.

13. See Gretchen Ruethling, "In Minnesota, Flu Vaccines Go Waiting," New York Times, November 12, 2004, www.nytimes.com/2004/11/12/national/12flu .html.

14. For details on connected individualism see Wellman (2002) and Rainie and Wellman (2012).

15. Detailed reports on the impact of technology on well-being in different communities, including the northern Ontario town of Chapleau, are accessible at "Cyber Society Publications," NETLAB, http://groups.chass.utoronto.ca /netlab/publications/cyber-society.

16. I draw here on the dissertations of Sanna Balsari-Palsule (2011) and Jean Arlt (2011).

17. Details of this are contained in a report to the Social Sciences and Humanities Research Council of Canada (Little, 1988).

18. See Misra and Stokols (2012).

NOTES TO CHAPTER 9

1. I introduced personal projects as ways of looking at personality in Little (1983). Among the early publications on the method were Palys and Little (1983) and Little (1989). The most comprehensive source for research on personal projects is Little, Salmela-Aro, and Phillips (2007).

2. On a technical note, we need to make some differentiations between actions, intentions, and projects. Action has an intentional aspect to it, whereas

behavior does not have this requirement. Consider the rapid closing and unclosing of one's eyes: this could be a mere behavior, without intentionality, as occurs in "blinking." But it could equally be seen, at least if done with one eye, as winking. The one is reflex behavior; the other is purposeful action. Or, more subtly, consider an optometrist who is instructing a patient on using eye drops effectively. When she demonstrates how to blink properly, that is action, not reflex behavior.

3. For details, see Little and Gee (2007).

4. See Nancy Cantor (1990) on the "having and doing" aspects of personality.

5. See Henrich, Heine, and Norenzayan (2010).

6. See Chambers (2007).

7. See Little and Gee (2007) and Little and Coulombe (in press).

8. These data were originally reported in Little (1987a).

9. For reviews of empirical studies on the relative importance of personal project dimensions for predicting well-being, see Little, Salmela-Aro, and Phillips (2007).

10. I discuss this in more detail in Little (1998).

11. A convincing case for the joint importance of meaning and efficacy in project pursuit is found in the work of Sheldon and Kasser (1998).

12. See his compellingly readable *Stumbling on Happiness* (Gilbert, 2006).

13. The most comprehensive review of studies in this area is that of Salmela-Aro and Little (2007).

14. See Hwang (2004).

15. See the fascinating research of Taylor, Klein, Lewis, Gruenewald, Gurung, and Updegraff (2000) on the tend-and-befriend response.

16. See Phillips, Little, and Goodine (1997).

17. See McKeen (1984).

18. See Dowden (2004).

NOTES TO CHAPTER 10

1. For details on the Golden Apple award and its guiding philosophy, see "Golden Apple Award: The University of Michigan Golden Apple Award," University of Michigan, 2014, http://goldenappleumich.wordpress.com.

2. An excellent examination of the relation between core projects and resistance to change appears in McDiarmid (1990).

3. See their chapters and commentaries in Stablein and Frost (2004).

4. See Crum and Langer (2007).

5. This technique, known as concept matching, has long been used in organizations to stimulate creative solutions to "stuck" problems. See Osborn (1953).

6. The most comprehensive account of Fixed Role Therapy appears in Kelly (1955). See also Epting and Nazario (1987).

7. I have adopted the "butler" example from a thoughtful article on certain paradoxical aspects of fixed role therapy by Han Bonarius (1970), one of the early pioneers in the field of personal construct psychology.

8. This example draws from the valuable web resource on personal construct therapy by Boeree (2006).

9. See McGregor, McAdams, and Little (2006).

10. Research on such counter-dispositional behavior is currently a hot area of inquiry. See, for example, Fleeson, Malanos, and Achille (2002), Whelan (2013), and Zelenski, Santoro, and Whelan (2012).

11. See Salmela-Aro (1992) and Little (1989).

12. See Melia-Gordon (1994).

13. Research on self-determination theory is burgeoning. For an overview of the theory, see Deci and Ryan (2002).

14. See Sulloway (1996).

15. Valerie Tiberius (2008) has written an evocative and insightful analysis of how such considerations emerge in the process of reflecting on our lives. Her treatment of personal projects in the context of such reflection is a compelling one.

16. The notion of reconciliation is closely related to recent research on the need for self-compassion, which has been shown to have a more salutary effect on individuals than self-esteem. See Leary, Tate, Adams, Allen, and Hancock (2007) and Neff (2003).

17. This is taken from the last chapter of Flanagan's intriguing book *Self Expressions: Mind, Morals, and the Meaning of Life* (1996).

References

Alexander, C. (1964). *Notes on the synthesis of form.* Cambridge, MA: Harvard University Press.

Alexander, C. (1970). The city as a mechanism for sustaining human contact. In W. Ewald (Ed.), *Environment for man* (pp. 60–102). Bloomington: Indiana University Press.

Antonovsky, A. (1979). *Health, stress, and coping.* San Francisco: Jossey-Bass.

Argyle, M., & Little, B. R. (1972). Do personality traits apply to social behaviour? *Journal for the Theory of Social Behaviour, 2*(1), 1–33.

Arlt, J. (2011). *Human contact and well-being: Exploring emotional intimacy and successful project pursuit on Facebook.* Unpublished master's thesis, Cambridge University, Cambridge, UK.

Asch, S. E. (1940). Studies in the principles of judgments and attitudes: II. Determination of judgments by group and by ego standards. *Journal of Social Psychology, 12*(2), 433–465.

Bagg, C. E., & Crookes, T. G. (1975). The responses of neonates to noise, in relation to the personalities of their parents. *Developmental Medicine & Child Neurology, 17*(6), 732–735.

Balsari-Palsule, S. (2011). *Human connection, personal projects and social networking sites.* Unpublished master's thesis, Cambridge University, Cambridge, UK.

Barefoot, J. C., & Boyle, S. H. (2009). Hostility and proneness to anger. In M. R. Leary & R. H. Hoyle (Eds.), *Handbook of individual differences in social behavior* (pp. 210–226). New York: Guilford.

Barnes, G. E. (1975). Extraversion and pain. *British Journal of Social and Clinical Psychology, 14*(3), 303–308.

Barrick, M. R., & Mount, M. K. (1991). The Big Five personality dimensions and job performance: A meta-analysis. *Personnel Psychology, 44*(1), 1–26.

Barron, F. (1953). An ego-strength scale which predicts response to psychotherapy. *Journal of Consulting Psychology, 17*(5), 327–333.

Barron, F. (1963). *Creativity and psychological health: Origins of personal vitality and creative freedom.* Princeton, NJ: Van Nostrand.

Bem, S. (1974). The measurement of psychological androgyny. *Journal of Consulting and Clinical Psychology, 42*(2), 155–162.

Berra, T. M. (2013). *Darwin and his children: His other legacy.* New York: Oxford University Press.

Biondo, J., & MacDonald, A. P. (1971). Internal-external locus of control and response to influence attempts. *Journal of Personality, 39*, 407–419.

Boeree, C. G. (2006). George Kelly. http://webspace.ship.edu/cgboer/kelly.html.

Bogg, T., & Roberts, B. W. (2004). Conscientiousness and health-related behaviors: A meta-analysis of the leading behavioral contributors to mortality. *Psychological Bulletin, 130*(6), 887–919.

Bonarius, J. C. J. (1970). Fixed role therapy: A double paradox. *British Journal of Medical Psychology, 43*(3), 213–219.

Booth-Kewley, S., & Vickers, R. R. Jr. (1994). Associations between major domains of personality and health behavior. *Journal of Personality, 62*(3), 281–298.

Buss, D. M. (1991). Evolutionary personality psychology. *Annual Review of Psychology, 42*, 459–491.

Buss, D. M. (2008). *Evolutionary psychology: The new science of the mind.* Boston: Pearson.

Cain, S. (2012). *Quiet: The power of introverts in a world that can't stop talking.* New York: Crown.

Cantor, N. (1990). From thought to behavior: "Having" and "doing" in the study of personality and cognition. *American Psychologist, 45*(6), 735–750.

Cantor, N., Norem, J. K., Niedenthal, P. M., Langston, C. A., and Brower, A. M. (1987). Life tasks, self-concept ideals, and cognitive strategies in a life transition. *Journal of Personality and Social Psychology, 53*(6), 1178–1191.

Carson, S. H., Peterson, J. B., & Higgins, D. M. (2003). Decreased latent inhibition is associated with increased creative achievement in high-functioning individuals. *Journal of Personality and Social Psychology, 85*(3), 499–506.

Carver, C. S., & Scheier, M. F. (1992). *Perspectives on personality* (2nd ed.). Boston: Allyn and Bacon.

Casey, B. J., Somerville, L. H., Gotlib, I. H., Ayduk, O., Franklin, N. T., Askren,

M. K., et al. (2011). Behavioral and neural correlates of delay of gratification 40 years later. *Proceedings of the National Academy of Sciences, 108*(36), 14998–15003.

Caspi, A. & Moffitt, T. E. (1993). When do individual differences matter? A paradoxical theory of personality coherence. *Psychological Inquiry, 4*(4), 247–271.

Chambers, N. C. (2007). Just doing it: Affective implications of project phrasing. In B. R. Little, K. Salmela-Aro, & S. D. Phillips (Eds.), *Personal project pursuit: Goals, action, and human flourishing* (pp. 145–169). Mahwah, NJ: Lawrence Erlbaum.

Charles, S. T., Gatz, M., Kato, K., & Pedersen, N. L. (2008). Physical health 25 years later: The predictive ability of neuroticism. *Health Psychology, 27*(3), 369–378.

Coleman, J. S., Campbell, E. Q., Hobson, C. J., McPartland, J., Mood, A. M., Weinfeld, F. D., & York, R. L. (1966). *Equality of educational opportunity.* Washington, DC: US Department of Health, Education, and Welfare Office of Education.

Costa, P. T., & McCrae, R. R. (1992). *NEO PI-R professional manual.* Odessa, FL: Psychological Assessment Resources.

Crowne, D. P., & Liverant, S. (1963). Conformity under varying conditions of personal commitment. *Journal of Abnormal and Social Psychology, 66*(6), 547–555.

Crum, A. J., & Langer, E. J. (2007). Mind-set matters: Exercise and the placebo effect. *Psychological Science, 18*(2), 165–171.

Deci, E. L., & Ryan, R. M. (2002). Self-determination research: Reflections and future directions. In E. L. Deci & R. M. Ryan (Eds.), *Handbook of self-determination research* (pp. 431–441). Rochester, NY: University of Rochester Press.

DeYoung, C. G. (2010). Personality neuroscience and the biology of traits. *Social and Personality Psychology Compass, 4*(12), 1165–1180.

Dowden, C. (2004). *Managing to be "free": Personality, personal projects and well-being in entrepreneurs.* Unpublished doctoral dissertation, Carleton University, Ottawa, Canada.

Dumont, F. (2010). *A history of personality psychology: Theory, science, and research from Hellenism to the twenty-first century.* New York: Cambridge University Press.

Duncan, J., Jackson, R., Lance, C. E., & Hoffman, B. J. (Eds.). (2012). *The psychology of assessment centers.* New York: Routledge.

Elliott, C. D. (1971). Noise tolerance and extraversion in children. *British Journal of Psychology, 62*(3), 375–380.

Endler, N. S., & Magnusson, D. (1976). Toward an interactional psychology of personality. *Psychological Bulletin, 83*(5), 956–974.

Epting, F. R., & Nazario, A., Jr. (1987). Designing a fixed role therapy: Issues, technique, and modifications. In R. A. Neimeyer & G. J. Neimeyer (Eds.), *Personal construct psychotherapy casebook* (pp. 277–289). New York: Springer.

Eysenck, H. J. (1967). *The biological basis of personality*. Springfield, IL: Thomas.

Eysenck, S. B. G., & Eysenck, H. J. (1967). Salivary response to lemon juice as a measure of introversion. *Perceptual and Motor Skills, 24*(3c), 1047–1053.

Faulkner, R. R., & Becker, H. S. (2009). *"Do you know . . . ?": The jazz repertoire in action*. Chicago: University of Chicago Press.

Flanagan, O. (1996). *Self expressions: Mind, morals, and the meaning of life*. New York: Oxford University Press.

Fleeson, W., Malanos, A. B., & Achille, N. M. (2002). An intraindividual process approach to the relationship between extraversion and positive affect: Is acting extraverted as "good" as being extraverted? *Journal of Personality and Social Psychology, 83*(6), 1409–1422.

Florida, R. (2008). *Who's your city? How the creative economy is making where to live the most important decision of your life*. Toronto: Random House of Canada.

Fransella, F. (Ed.). (2003). *International handbook of personal construct psychology*. Chichester, UK: Wiley.

Friedman, H. S., Tucker, J. S., Tomlinson-Keasey, C., Schwartz, J. E., Wingard, D. L., & Criqui, M. H. (1993). Does childhood personality predict longevity? *Journal of Personality and Social Psychology, 65*(1), 176–185.

Furnham, A. (2009). Locus of control and attribution style. In M. R. Leary & R. H. Hoyle (Eds.), *Handbook of individual differences in social behaviour* (pp. 274–287). New York: Guilford.

Gilbert, D. (2006). *Stumbling on happiness*. New York: Alfred A. Knopf.

Glass, D. C., & Singer, J. E. (1972). Behavioral aftereffects of unpredictable and uncontrollable aversive events. *American Scientist, 60*(4), 457–465.

Gollwitzer, P. M., & Kinney, R. F. (1989). Effects of deliberative and implemental mind-sets on illusion of control. *Journal of Personality and Social Psychology, 56*(4), 531–542.

Goncalo, J. A., Flynn, F. J., & Kim, S. H. (2010). Are two narcissists better than one? The link between narcissism, perceived creativity, and creative performance. *Personality and Social Psychology Bulletin, 36*(11), 1484–1495.

Gosling, S. D. (2009, September/October). Mixed signals. *Psychology Today, 42*(5), 62–71.

Gosling, S. D., Rentfrow, P. J., & Swann Jr, W. B. (2003). A very brief measure of the Big-Five personality domains. *Journal of Research in Personality, 37*(6), 504–528.

Gough, H. G. (1979). A creative personality scale for the Adjective Check List. *Journal of Personality and Social Psychology, 37*(8), 1398–1405.

Grant, A. M. (2013). Rethinking the extraverted sales ideal: The ambivert advantage. *Psychological Science, 24*(6), 1024–1030.

Graziano, W. G., & Tobin, R. M. (2008). Agreeableness. In M. R. Leary & R. H. Hoyle (Eds.), *Handbook of individual differences in social behavior* (pp. 46–61). New York: Guilford.

Hass, R. G. (1984). Perspective-taking and self-awareness: Drawing an E on your forehead. *Journal of Personality and Social Psychology, 46*(4), 788–798.

Helson, R. (1971). Women mathematicians and the creative personality. *Journal of Consulting and Clinical Psychology, 36*(2), 210–220.

Hennessey B. A., & Amabile, T. M. (1998). Reality, intrinsic motivation, and creativity. *American Psychologist, 53*(6), 674–675.

Henrich, J., Heine, S. J., & Norenzayan, A. (2010). The weirdest people in the world? *Behavioral and Brain Sciences, 33*(2–3), 61–83.

Hinkle, D. N. (1965). *The change of personal constructs from the viewpoint of a theory of construct implications.* Unpublished doctoral dissertation, Ohio State University, Columbus, OH.

Hochschild, A. R. (1983). *The managed heart: The commercialization of human feeling.* Berkeley: University of California Press.

Hogan, J., & Hogan, R. (1993). *Ambiguities of conscientiousness.* Paper presented at the 8th Annual Conference of the Society for Industrial and Organizational Psychology, San Francisco, CA.

Holmes, T. H., & Rahe, R. H. (1967). The social readjustment rating scale. *Journal of Psychosomatic Research, 11*(2), 213–218.

Howarth, E., & Skinner, N. F. (1969). Salivation as a physiological indicator of introversion. *Journal of Psychology, 73*(2), 223–228.

Hwang, A. A. (2004). *Yours, mine, ours: The role of joint personal projects in close relationships.* Unpublished doctoral dissertation, Harvard University, Cambridge, MA.

James, W. (1902). *The varieties of religious experience.* London: Longmans, Green.

Jang, K. L., Livesley, W. J., & Vernon, P. A. (1996). Heritability of the Big Five personality dimensions and their facets: A twin study. *Journal of Personality, 64*(3), 577–591.

John, O. P., Naumann, L. P., & Soto, C. J. (2008). Paradigm shift to the integrative Big-Five Trait Taxonomy: History, measurement, and conceptual issues. In O. P. John, R. W. Robins, & L. A. Pervin (Eds.), *Handbook of personality: Theory and research* (pp. 114–158). New York: Guilford Press.

Jönsson, P., & Carlsson, I. (2000). Androgyny and creativity. *Scandinavian Journal of Psychology, 41*(4), 269–274.

Judge, T. A., Livingston, B. A., & Hurst, C. (2012). Do nice guys—and gals—really finish last? The joint effects of sex and agreeableness on income. *Journal of Personality and Social Psychology, 102*(2), 390–407.

Jung, C. G. (1921). *Psychological types.* Princeton, NJ: Princeton University Press.

Kelley, H. H., & Michela, J. L. (1980). Attribution theory and research. *Annual Review of Psychology, 31,* 457–501.

Kelly, G. A. (1955). *The psychology of personal constructs.* New York: Norton.

Kelly, G. A. (1958). Man's construction of his alternatives. In G. Lindzey (Ed.), *The assessment of human motives* (pp. 33–64). New York: Van Nostrand.

Kilduff, M., & Day, D. V. (1994). Do chameleons get ahead? The effects of self-monitoring on managerial careers. *Academy of Management Journal, 37*(4), 1047–1060.

Kluckhohn, C., & Murray, H. A. (Eds.). (1953). *Personality in nature, society and culture.* New York: Knopf.

Kogan, A., Saslow, L. R., Impett, E. A., Oveis, C., Keltner, D., & Saturn, S. R. (2011). Thin-slicing study of the oxytocin receptor (OXTR) gene and the evaluation and expression of the prosocial disposition. *Proceedings for the National Academy of Sciences, 108*(48), 19189–19192.

Lambert, C. (2003, July/August). Traits of Gibraltar? Introversion unbound. *Harvard Magazine,* 12–14.

Langer, E. J., & Rodin, J. (1976). The effects of choice and enhanced personal responsibility for the aged: A field experiment in an institutional setting. *Journal of Personality and Social Psychology, 34*(2), 191–198.

Leary, M. R., Tate, E. B., Adams, C. E., Allen, A. B., & Hancock, J. (2007). Self-compassion and reactions to unpleasant self-relevant events: The implications of treating oneself kindly. *Journal of Personality and Social Psychology, 92*(5), 887–904.

Lefcourt, H. M. (1982). *Locus of control: Current trends in theory and research* (2nd ed.). Hillsdale, NJ: Lawrence Erlbaum.

Leone, C., & Hawkins, L. B. (2006). Self-monitoring and close relationships. *Journal of Personality, 74*(3), 739–778.

Lester, D. (2009). Emotions in personal construct theory: A review. *Personal Construct Theory & Practice, 6,* 90–98.

Levine, R. V., Lynch, K., Miyake, K., & Lucia, M. (1989). The Type A city: Coronary heart disease and the pace of life. *Journal of Behavioral Medicine*, *12*(6), 509–524.

Little, B. R. (1972). Psychological man as scientist, humanist and specialist. *Journal of Experimental Research in Personality*, *6*, 95–118.

Little, B. R. (1976). Specialization and the varieties of environmental experience: Empirical studies within the personality paradigm. In S. Wapner, S. B. Cohen, & B. Kaplan (Eds.), *Experiencing the environment* (pp. 81–116). New York: Plenum.

Little, B. R. (1983). Personal projects: A rationale and method for investigation. *Environment and Behavior*, *15*(3), 273–309.

Little, B. R. (1987a). Personal projects and fuzzy selves: Aspects of self-identity in adolescence. In T. Honess & K. Yardley (Eds.), *Self and identity: Perspectives across the lifespan* (pp. 230–245). New York: Routledge.

Little, B. R. (1987b). Personality and the environment. In D. Stokols & I. Altman (Eds.), *Handbook of environmental psychology* (pp. 205–244). New York: Wiley.

Little, B. R. (1988). *Personal projects analysis: Theory, method and research*. Final report to the Social Sciences and Humanities Research Council of Canada. Social Ecological Laboratory, Carleton University, Ottawa, Canada.

Little, B. R. (1989). Personal projects analysis: Trivial pursuits, magnificent obsessions, and the search for coherence. In D. Buss & N. Cantor (Eds.), *Personality psychology: Recent trends and emerging directions* (pp. 15–31). New York: Springer-Verlag.

Little, B. R. (1996). Free traits, personal projects and idio-tapes: Three tiers for personality psychology. *Psychological Inquiry*, *7*(4), 340–344.

Little, B. R. (1998). Personal project pursuit: Dimensions and dynamics of personal meaning. In P. T. P. Wong & P. S. Fry (Eds.), *The human quest for meaning: A handbook of psychological research and clinical applications* (pp. 197–221). Thousand Oaks, CA: Sage.

Little, B. R. (1999a). Personal projects and social ecology: Themes and variation across the life span. In J. Brandtstadter & R. M. Lerner (Eds.), *Action and self-development: Theory and research through the life span* (pp. 197–221). Thousand Oaks, CA: Sage.

Little, B. R. (1999b). Personality and motivation: Personal action and the conative evolution. In L. A. Pervin & O. P. John (Eds.), *Handbook of personality theory and research* (2nd ed., pp. 501–524). New York: Guilford.

Little, B. R. (2000). Free traits and personal contexts: Expanding a social ecological model of well-being. In. W. B. Walsh, K. H. Craik, & R. H. Price (Eds.), *Person-environment psychology: New directions and perspectives* (2nd ed., pp. 87–116). Mahwah, NJ: Lawrence Erlbaum.

Little, B. R. (2005). Personality science and personal projects: Six impossible things before breakfast. *Journal of Research in Personality, 39*, 4–21.

Little, B. R. (2007). Prompt and circumstance: The generative contexts of personal projects analysis. In B. R. Little, K. Salmela-Aro, & S. D. Phillips (Eds.), *Personal project pursuit: Goals, action, and human flourishing* (pp. 3–49). Mahwah, NJ: Lawrence Erlbaum.

Little, B. R. (2010). Opening space for project pursuit: Affordance, restoration and chills. In C. W. Thompson, P. Aspinall, & S. Bell (Eds.), *Innovative approaches to researching landscape and health. Open space: People space 2* (pp. 163–178). New York: Routledge.

Little, B. R. (2011). Personality science and the northern tilt: As positive as possible under the circumstances. In K. M. Sheldon, T. B. Kashdan, & M. F. Steger (Eds.), *Designing positive psychology: Taking stock and moving forward* (pp. 228–247). New York: Oxford University Press.

Little, B. R., & Coulombe, S. (in press). Personal projects analysis. In *International encyclopedia of social and behavioral sciences* (2nd ed.). Oxford, UK: Elsevier.

Little, B. R., & Gee, T. L. (2007). The methodology of personal projects analysis: Four modules and a funnel. In B. R. Little, K. Salmela-Aro, & S. D. Phillips (Eds.), *Personal project pursuit: Goals, action, and human flourishing* (pp. 51–94). Mahwah, NJ: Lawrence Erlbaum.

Little, B. R., & Joseph, M. F. (2007). Personal projects and free traits: Mutable selves and well beings. In B. R. Little, K. Salmela-Aro, & S. D. Phillips (Eds.), *Personal project pursuit: Goals, action, and human flourishing* (pp. 375–400). Mahwah, NJ: Lawrence Erlbaum.

Little, B. R., Salmela-Aro, K., & Phillips, S. D. (Eds.). (2007). *Personal project pursuit: Goals, action, and human flourishing*. Mahwah, NJ: Lawrence Erlbaum.

Loo, R. (1979). Role of primary personality factors in the perception of traffic signs and driver violations and accidents. *Accident Analysis and Prevention, 11*(2), 125–127.

Lynn, R., & Eysenck, H. J. (1961). Tolerance for pain, extraversion and neuroticism. *Perceptual and Motor Skills, 12*(2), 161–162.

MacDonald, A. P. (1970). Internal-external locus of control and the practice of birth control. *Psychological Reports, 27,* 206.

MacKinnon, D. W. (1962). The nature and nurture of creative talent. *American Psychologist, 17*(7), 484–495.

MacKinnon, D. W. (1965). Personality and the realization of creative potential. *American Psychologist, 20*(4), 273–281.

Maddi, S. R., & Kobasa, S. C. (1984). *The hardy executive: Health under stress.* Homewood, IL: Dow Jones-Irwin.

Mahlamaki, T. (2010). *Influence of personality on the job performance of key account managers.* Unpublished doctoral dissertation, Tampere University of Technology, Tampere, Finland.

McAdams, D. P. (1995). What do we know when we know a person? *Journal of Personality, 63*(3), 365–396.

McAdams, D. P. (2009). *The person: An introduction to the science of personality psychology* (5th ed.). Hoboken, NJ: Wiley.

McAdams, D. P. (2010). *George W. Bush and the redemptive dream: A psychological portrait.* New York: Oxford University Press.

McCrae, R. R. (2007). Aesthetic chills as a universal marker of openness to experience. *Motivation and Emotion, 31*(1), 5–11.

McCrae, R. R., & Sutin, A. R. (2009). Openness to experience. In M. R. Leary & R. H. Hoyle (Eds.), *Handbook of individual differences in social behavior* (pp. 257–273). New York: Guilford.

McDiarmid, E. (1990). *Level of molarity, project cross-impact and resistance to change in personal project systems.* Unpublished master's thesis, Carleton University, Ottawa, Canada.

McGregor, I., McAdams, D. P., & Little, B. R. (2006). Personal projects, life stories, and happiness: On being true to traits. *Journal of Research in Personality, 40*(5), 551–572.

McKechnie, G. E. (1977). The Environmental Response Inventory in application. *Environment and Behavior, 9*(2), 255–276.

McKeen, N. A. (1984). *The personal projects of pregnant women.* Unpublished bachelor's thesis, Carleton University, Ottawa, Canada.

Melia-Gordon, M. (1994). *The measurement and meaning of personal projects creativity.* Unpublished master's thesis, Carleton University, Ottawa, Canada.

Milgram, S. (1970). The experience of living in cities. *Science, 167*(3924), 1461–1468.

Mischel, W. (1968). *Personality and assessment.* New York: Wiley.

Mischel, W., Ebbesen, E. B., & Zeiss, A. R. (1972). Cognitive and attentional mechanisms in delay of gratification. *Journal of Personality and Social Psychology*, *21*(2), 204–218.

Misra, S., & Stokols, D. (2012). Psychological and health outcomes of perceived information overload. *Environment and Behavior*, *44*(6), 737–759.

Moskowitz, D. S., & Coté, S. (1995). Do interpersonal traits predict affect? A comparison of three models. *Journal of Personality and Social Psychology*, *69*(5), 915–924.

Murray, H. A. (1938). *Explorations in personality*. New York: Oxford University Press.

Myers, I. B., McCaulley, M. H., Quenk, N. L., & Hammer, A. L. (1998). *MBTI Manual: A guide to the development and use of the Myers-Briggs Type Indicator* (3rd ed.). Palo Alto, CA: Consulting Psychologists Press.

Neff, K. D. (2003). The development and validation of a scale to measure self-compassion. *Self and Identity*, *2*(3), 223–250.

Nettle, D. (2006). The evolution of personality variation in humans and other animals. *American Psychologist*, *61*(6), 622–631.

Nettle, D. (2007). *Personality: What makes you the way you are*. New York: Oxford University Press.

Ng, T. W. H., Sorensen, K. L., & Eby, L. T. (2006). Locus of control at work: A meta-analysis. *Journal of Organizational Behavior*, *27*(8), 1057–1087.

Norem, J. K. (2002). *The power of negative thinking: Using defensive pessimism to manage anxiety and perform at your peak*. New York: Basic Books.

Osborn, A. F. (1953). *Applied imagination: Principles and procedures of creative problem-solving*. New York: Charles Scribner's Sons.

Ozer, D. J., & Benet-Martínez, V. (2006). Personality and the prediction of consequential outcomes. *Annual Review of Psychology*, 57, 401–421.

Palys, T. S., & Little, B. R. (1983). Perceived life satisfaction and the organization of personal project systems. *Journal of Personality and Social Psychology*, *44*(6), 1221–1230.

Paul, A. M. (2004). *The cult of personality*. New York: Free Press.

Paulhus, D. L. (1983). Sphere-specific measures of perceived control. *Journal of Personality and Social Psychology*, *44*(6), 1253–1265.

Paulhus, D. L., & Martin, C. L. (1987). The structure of personality capabilities. *Journal of Personality and Social Psychology*, *52*(2), 354–365.

Pennebaker, J. W. (1990). *Opening up: The healing power of expressing emotions*. New York: Guilford.

Pennebaker, J. W., Kiecolt-Glaser, J. K., & Glaser, R. (1988). Disclosure of traumas and immune function: Health implications for psychotherapy. *Journal of Consulting and Clinical Psychology, 56*(2), 239–245.

Peterson, J. B., Smith, K. W., & Carson, S. (2002). Openness and extraversion are associated with reduced latent inhibition: Replication and commentary. *Personality and Individual Differences, 33*(7), 1137–1147.

Phares, E. J. (1965). Internal-external control as a determinant of amount of social influence exerted. *Journal of Personality and Social Psychology, 2*(5), 642–647.

Phillips, S. D., Little, B. R., & Goodine, L. A. (1997). Reconsidering gender and public administration: Five steps beyond conventional research. *Canadian Public Administration, 40*(4), 563–581.

Pickering, G. W. (1974). *Creative malady: Illness in the lives and minds of Charles Darwin, Florence Nightingale, Mary Baker Eddy, Sigmund Freud, Marcel Proust and Elizabeth Barrett Browning*. New York: Dell.

Pittenger, D. J. (1993). Measuring the MBTI . . . and coming up short. *Journal of Career Planning and Employment, 54*(1), 48–52.

Platt, E. S. (1969). Internal/external control and changes in expected utility as predictors of change in cigarette smoking following role playing. Paper presented at Eastern Psychological Association Convention, Philadelphia, PA.

Rainie, L., & Wellman, B. (2012). *Networked: The new social operating system.* Cambridge: Massachusetts Institute of Technology.

Rentfrow, P. J., Gosling, S. D., & Potter, J. (2008). A theory of the emergence, persistence, and expression of geographic variation in psychological characteristics. *Perspectives on Psychological Science, 3*(5), 339–369.

Revelle, W., Humphreys, M. S., Simon, L., & Gilliland, K. (1980). The interactive effect of personality, time of day, and caffeine: A test of the arousal model. *Journal of Experimental Psychology: General, 109*(1), 1–39.

Roberts, B. W., & DelVecchio, W. F. (2000). The rank-order consistency of personality traits from childhood to old age: A quantitative review of longitudinal studies. *Psychological Bulletin, 126*(1), 3–25.

Roberts, B. W., & Robins, R. W. (2003). Person-environment fit and its implications for personality development: A longitudinal study. *Journal of Personality, 72*(1), 89–110.

Rotter, J. B. (1966). Generalized expectancies for internal versus external control of reinforcement. *Psychological Monographs: General and Applied, 80*(1), 1–28.

Ryle, A. (1975). *Frames and cages: The repertory grid approach to human understanding.* Oxford, UK: International Universities Press.

Salmela-Aro, K. (1992). Struggling with self: The personal projects of students seeking psychological counselling. *Scandinavian Journal of Psychology*, *33*(4), 330–338.

Salmela-Aro, K., & Little, B. R. (2007). Relational aspects of project pursuit. In B. R. Little, K. Salmela-Aro, & S. D. Phillips (Eds.), *Personal project pursuit: Goals, action, and human flourishing* (pp. 199–219). Mahwah, NJ: Lawrence Erlbaum.

Samuels, D. B., & Widiger, T. A. (2011). Conscientiousness and obsessive-compulsive personality disorder. *Personality Disorders: Theory, Research and Treatment* 2(3), 161–174.

Scheibe, K. E. (2010). The person as actor, the actor as person: Personality from a dramaturgical perspective. *Psicologia da Educação*, (31), 65–78.

Schmitt, D. P., Allik, J., McCrae, R. R., Benet-Martínez, V., Alcalay, L., Ault, L. et al. (2007). The geographic distribution of Big Five personality traits: Patterns and profiles of human self-description across 56 nations. *Journal of Cross-Cultural Psychology*, *38*(2), 173–212.

Schulz, R., & Hanusa, B. H. (1978). Long-term effects of control and predictability-enhancing interventions: Findings and ethical issues. *Journal of Personality and Social Psychology*, *36*(11), 1194–1201.

Seeman, M. (1963). Alienation and social learning in a reformatory. *American Journal of Sociology*, *69*(3), 270–284.

Seligman, M. E. P. (2011). *Flourishing: A visionary new understanding of happiness and well-being*. New York: Free Press.

Sheldon, K. M., & Kasser, T. (1998). Pursuing personal goals: Skills enable progress, but not all progress is beneficial. *Personality and Social Psychology Bulletin*, *24*(12), 1319–1331.

Smith, A. P. (2013). Caffeine, extraversion and working memory. *Journal of Psychopharmacology*, *27*(1), 71–76.

Snyder, M. (1974). Self-monitoring of expressive behavior. *Journal of Personality and Social Psychology*, *30*(4), 526–537.

Snyder, M. (1979). Self-monitoring processes. In L. Berkowitz (Ed.), *Advances in experimental social psychology* (Vol. 12, pp. 85–128). New York: Academic Press.

Snyder, M. (1987). *Public appearances, private realities: The psychology of self-monitoring*. New York: W. H. Freeman.

Snyder, M., & Gangestad, S. (1986). Choosing social situations: Two investigations of self-monitoring processes. *Journal of Personality and Social Psychology*, *43*(1), 123–135.

Snyder, M., & Gangestad, S. (1986). On the nature of self-monitoring: Matters of assessment, matters of validity. *Journal of Personality and Social Psychology, 51*(1), 125–139.

Snyder, M., Gangestad, S., & Simpson, J. A. (1983). Choosing friends as activity partners: The role of self-monitoring. *Journal of Personality and Social Psychology, 45*(5), 1061–1072.

Snyder, M., & Simpson, J. A. (1984). Self-monitoring and dating relationships. *Journal of Personality and Social Psychology, 47*(6), 1281–1291.

Snyder, M., & Simpson, J. A. (1987). Orientations toward romantic relationships. In D. Perlman & S. Duck (Eds.), *Intimate relationships: Development, dynamics, and deterioration* (pp. 45–62). Newbury Park, CA: Sage.

Stablein, R. E., & Frost, P. J. (Eds.). (2004). *Renewing research practice*. Stanford, CA: Stanford University Press.

Steel, P., Schmidt, J., & Shultz, J. (2008). Refining the relationship between personality and subjective well-being. *Psychological Bulletin, 134*(1), 138–161.

Sulloway, F. J. (1996). *Born to rebel: Birth order, family dynamics, and creative lives*. New York: Pantheon.

Taylor, S. E., & Brown, J. D. (1988). Illusion and well-being: A social psychological perspective on mental health. *Psychological Bulletin, 103*(2), 193–210.

Taylor, S. E., Klein, L. C., Lewis, B. P., Gruenewald, T. L., Gurung, R.A.R., & Updegraff, J. A. (2000). Biobehavioral responses to stress in females: Tend-and-befriend, not fight-or-flight. *Psychological Review, 107*(3), 411–429.

Tiberius, V. (2008). *The reflective life: Living with our limits*. Oxford, UK: Oxford University Press.

Triandis, H. C., & Suh, E. M. (2002). Cultural influences on personality. *Annual Review of Psychology, 53*, 133–160.

Turner, R. G. (1980). Self-monitoring and humor production. *Journal of Personality, 48*(2), 163–167.

Vallacher, R. R., & Wegner, D. M. (1987). What do people think they're doing? Action identification and human behavior. *Psychological Review, 94*(1), 3–15.

Vinokur, A., & Selzer, M. L. (1975). Desirable versus undesirable life events: Their relationship to stress and mental distress. *Journal of Personality and Social Psychology, 32*(2), 329–337.

von Knorring, L., von Knorring, A.-L., Mornstad, H., & Nordlund, Å. (1987). The risk of dental caries in extraverts. *Personality and Individual Differences, 8*(3), 343–346.

Walker, B. M., & Winter, D. A. (2007). The elaboration of personal construct psychology. *Annual Review of Psychology, 58*, 453–477.

Wallace, J. (1966). An abilities conception of personality: Some implications for personality measurement. *American Psychologist*, *21*(2), 132–138.

Weeks, D., & James, J. (1995). *Eccentrics: A study of sanity and strangeness*. New York: Villard.

Wegner, D. M. (1989). *White bears and other unwanted thoughts: Suppression, obsession, and the psychology of mental control*. New York: Viking/Penguin.

Wegner, D. M. (1994). Ironic processes of mental control. *Psychological Review*, *101*(1), 34–52.

Wellman, B. (2002). Little boxes, glocalization, and networked individualism. In M. Tanabe, P. van den Besselaar, & T. Ishida (Eds.), *Digital cities II: Computational and sociological approaches: Lecture notes in computer science* (vol. 2362, pp. 10–25). Berlin: Springer Berlin Heidelberg.

Whelan, D. C. (2013). *Extraversion and counter-dispositional behaviour: Exploring consequences and the impact of situation-behaviour congruence*. Unpublished doctoral dissertation, Carleton University, Ottawa, Canada.

Widiger, T. A. (2009). Neuroticism. In M. R. Leary and R. H. Hoyle (Eds.), *Handbook of individual differences in social behavior* (pp. 129–146). New York: Guildford.

Wilson, G. (1978). Introversion/extraversion. In H. London & J. E. Exner (Eds.), *Dimensions of personality* (pp. 217–261). New York: Wiley.

Wilt, J., & Revelle, W. (2009). Extraversion. In M. R. Leary & R. H. Hoyle (Eds.), *Handbook of individual differences in social behavior* (pp. 27–45). New York: Guilford.

Winter, D. G., & Barenbaum, N. B. (1999). History of modern personality theory and research. In L. A. Pervin & O. P. John (Eds.), *Handbook of personality: Theory and research* (2nd ed., pp. 3–27). New York: Guilford.

Zelenski, J. M., Santoro, M. S., & Whelan, D. C. (2012). Would introverts be better off if they acted more like extraverts? Exploring emotional and cognitive consequences of counterdispositional behavior. *Emotion*, *12*(2), 290–303.

Zemke, R. (1992). Second thoughts about the MBTI. *Training*, *29*(4), 43–47.

Index